REVISE EDEXCEL GCSE (9–1)
Biology

REVISION WORKBOOK

Higher

Series Consultant: Harry Smith
Author: Dr Stephen Hoare

A note from the publisher

In order to ensure that this resource offers high-quality support for the associated Pearson qualification, it has been through a review process by the awarding body. This process confirms that this resource fully covers the teaching and learning content of the specification or part of a specification at which it is aimed. It also confirms that it demonstrates an appropriate balance between the development of subject skills, knowledge and understanding, in addition to preparation for assessment.

Endorsement does not cover any guidance on assessment activities or processes (e.g. practice questions or advice on how to answer assessment questions), included in the resource nor does it prescribe any particular approach to the teaching or delivery of a related course.

While the publishers have made every attempt to ensure that advice on the qualification and its assessment

is accurate, the official specification and associated assessment guidance materials are the only authoritative source of information and should always be referred to for definitive guidance.

Pearson examiners have not contributed to any sections in this resource relevant to examination papers for which they have responsibility.

Examiners will not use endorsed resources as a source of material for any assessment set by Pearson.

Endorsement of a resource does not mean that the resource is required to achieve this Pearson qualification, nor does it mean that it is the only suitable material available to support the qualification, and any resource lists produced by the awarding body shall include this and other appropriate resources.

 Question difficulty
Look at this scale next to each exam-style question. It tells you how difficult the question is.

For the full range of Pearson revision titles across KS2, KS3, GCSE, Functional Skills, AS/A Level and BTEC visit:
www.pearsonschools.co.uk/revise

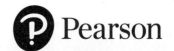

Contents

- - - - - - - - - - - - - - -

A small bit of small print:
Edexcel publishes Sample Assessment Material and the Specification on its website. This is the official content and this book should be used in conjunction with it. The questions have been written to help you practice every topic in the book. Remember: the real exam questions may not look like this.

Plant and animal cells

1 (a) Which of the following are found in both animal and plant cells?

☐ **A** cell membrane, nucleus, chloroplast

☐ **B** cell membrane, nucleus, ribosomes

☐ **C** cell wall, nucleus, ribosomes

☐ **D** cell wall, mitochondria, ribosomes   **(1 mark)**

> Look at the mark allocation for each question – here there is one mark so you need to put a cross in **one** box.

(b) Which of the following are found only in plant cells?

☐ **A** cell membrane, nucleus, chloroplast

☐ **B** cell membrane, vacuole, chloroplast

☐ **C** cell wall, chloroplast, vacuole

☐ **D** cytoplasm, chloroplast, vacuole   **(1 mark)**

> Always answer multiple-choice questions, even if you don't actually know the answer.

2 (a) Explain why muscle cells contain many mitochondria.

....

.... **(2 marks)**

(b) Explain why all plant cells contain mitochondria but only some contain chloroplasts.

> Chloroplasts need light to carry out photosynthesis. Use the function of a chloroplast to explain why you would not find them in certain cells, such as root cells.

....

....

....

.... **(2 marks)**

3 Describe the difference between the functions of a cell membrane and a cell wall.

> Guided >

Cell membrane controls ....

....

....

....

.... **(2 marks)**

4 The main function of fat cells is to store fat. Pancreatic exocrine cells secrete pancreatic juice, which contains many different digestive enzymes.

> Do not be put off by 'pancreatic exocrine cells'. Remember that enzymes are proteins and consider where in the cell proteins are made.

Suggest an explanation for why pancreatic exocrine cells contain many more ribosomes than fat cells.

....

....

.... **(2 marks)**

Different kinds of cell

> **Guided**

1 The genes in a bacterial cell are contained:

☐ **A** on a circular chromosome only

☐ **B** on plasmids only

☐ **C** on plasmids and a circular chromosome

☐ **D** ~~in the nucleus~~   It cannot be D because bacteria do not have nuclei.   **(1 mark)**

2 The diagram shows a sperm cell and a bacterium. Note that the drawings are not to the same scale.

Sperm cell   Bacterium

(a) Name the structures labelled A and B in the diagram:

A ....

B .... **(2 marks)**

(b) Describe the function of each structure.

A ....

....

B ....

.... **(2 marks)**

3 Breathing can expose us to dust, dirt and bacteria. Explain how cells in the lungs are specialised to protect us from these.

....

....

....

....

.... **(3 marks)**

4 The egg cell is much larger than the sperm cell. Give a reason to explain why.

....

.... **(2 marks)**

Microscopes and magnification

1 Scientists use two types of microscope to examine cells: light microscopes and electron microscopes. Describe how these types of microscope are different.

Light microscopes magnify......................than electron microscopes.

The level of cell detail seen with an electron microscope is ....

because .... **(3 marks)**

2 The image shows an electron micrograph of part of a human liver cell.

(a) Explain why this is a eukaryotic cell.

....

....

....**(2 marks)**

mitochondrion —

nucleus —

2 µm

(b) Estimate the size of the following parts of the cell:

(i) the nucleus

.... **(2 marks)**

(ii) the mitochondrion

.... **(2 marks)**

(c) Explain why it would be possible to see the nucleus clearly using a light microscope, but the mitochondria would be unclear.

....

....

.... **(3 marks)**

3 A scientist wants to study some bacteria that are 2.5 µm long. She can use either a light microscope (the one in the lab has a magnification of ×1000) or an electron microscope (the one in the lab next door has a magnification of ×100 000).

(a) Calculate the size of the magnified image of the bacteria seen with each type of microscope.

> Remember that $1\,µm = 1 \times 10^{-6}\,m$ and do a reality check on your answer. The magnified image must be **bigger** than the bacteria and the image formed by the electron microscope must be **bigger** than that formed by the light microscope.

(3 marks)

(b) Explain which microscope would be better for her to use.

....

....

.... **(2 marks)**

> State which is better **and** give a reason.

Dealing with numbers

Guided

1 Give the following units in order of increasing size:

metre   micrometre   millimetre   nanometre   picometre

picometre.. metre   **(1 mark)**

Guided

2 Complete the table to convert the quantities to the units shown:

Quantity	Converted quantity
0.005 nanometres	5 picometres
250 milligrams	grams
250 milligrams	kilograms
2.5 metres	millimetres

(4 marks)

Guided

3 For each of the following conversions, state whether it is true or false.

Conversion	True or false?
$0.000\,125\,mm = 0.125\,\mu m$	true
$150\,000\,mg = 0.015\,kg$	
$1\,kg = 10\,000\,000\,\mu g$	
$0.25\,mm = 2.5 \times 10^2\,\mu m$	

(4 marks)

4 Calculate for each of the following the actual size of the structure in metres (m) **in standard form** to **two significant figures**.

> You have to remember whether to multiply or divide (check back on page 3 of the Revision Guide) as well as get the standard form right **and** round to 2 significant figures (one place of decimals in standard form).

(a) a ribosome that measured 30.9 mm in an electron micrograph (magnification = ×1 000 000)

.... m   **(2 marks)**

(b) a mitochondrion that measured 163 mm in an electron micrograph (magnification = ×250 000)

.... m   **(2 marks)**

(c) a nucleus that measured 7.8 mm in a light microscope (magnification = ×800)

.... m   **(2 marks)**

 **Using a light microscope**

1 (a) State the function of the following parts of a light microscope.

(i) the mirror

.... **(1 mark)**

(ii) the stage with clips

.... **(1 mark)**

(iii) the coarse focusing wheel

.... **(1 mark)**

(b) Give the reasons for the following precautions when using a light microscope.

(i) Never use the coarse focusing wheel with a high power objective.

....

.... **(1 mark)**

(ii) Never point the mirror directly at the Sun.

....

.... **(1 mark)**

(c) (i) State an alternative light source that might be safer than the Sun.

.... **(1 mark)**

⟩ **Guided** ⟩

(ii) State **two** other precautions that you should take when using a light microscope.

precaution 1   Always start with the lowest power objective under the eyepiece.

precaution 2   ....

.... **(2 marks)**

2 You are observing a slide under high power but cannot see the part you need. Describe how you would bring the required part into view.

> Think about why you cannot see what you need and then the steps you must follow to find it. Remember some of the precautions you have to take.

....

....

....

....

....

.... **(3 marks)**

Practical skills Drawing labelled diagrams

1 A student was given the slide below left and told to make a high power drawing to show cells in different stages of mitosis. His drawing is shown below right.

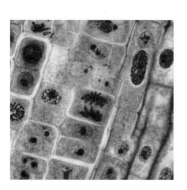

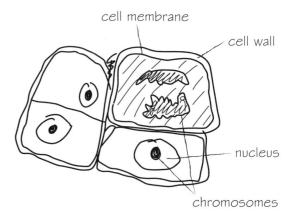

cell membrane

cell wall

nucleus

chromosomes

(a) Identify three faults with the student's drawing.

fault 1   The drawing is in pen rather than in ....

fault 2   ....

fault 3   ....   **(3 marks)**

(b) Draw your own labelled diagram of the slide above.

> Include outlines of all cells with more detail of cells showing different stages of mitosis. Try to show one of each stage.

(4 marks)

2 The student used a scale to measure the actual width of the field of view shown in the slide (above left) and found it was 0.113 mm. Calculate the magnification.

magnification = ...   **(3 marks)**

Enzymes

1   The enzyme invertase digests sucrose to glucose and fructose. Explain why invertase will not digest the sugar lactose.

Guided

The shape of .... matches the shape of

.... so .... cannot

combine with ....

.... **(2 marks)**

2   The graph shows how the rate of an enzyme reaction changes with temperature.

(a) Describe how the rate of reaction changes with temperature.

....

....

....

.... **(3 marks)**

(b) Explain the effect of temperature on the rate of reaction in the following areas of the curve.

> Relate your explanation to the shape of the curve. Consider the effect of temperature on protein structure.

(i) region A

....

.... **(2 marks)**

(ii) point B

....

.... **(2 marks)**

(iii) region C

....

.... **(2 marks)**

3   Pepsin and trypsin are proteases. Pepsin is produced in the stomach (pH 2), and trypsin is found in pancreatic juice (pH 8.6) released into the small intestine. Saliva and pancreatic juice both contain amylase. The graph shows the effect of pH on the activity of these enzymes.

Use this information to explain why proteins are digested in the stomach and small intestine, but starch is only digested in the mouth and small intestine.

....

....

....

....

....

....

.... **(4 marks)**

Practical skills

pH and enzyme activity

1 A student carried out an experiment to investigate the effect of pH on the activity of the enzyme trypsin using pieces of photographic film. Trypsin digests the protein in the film and causes the film to turn clear. Measuring the time it takes for the film to clear allows you to calculate the rate of reaction. The student used the apparatus shown.

This procedure was repeated using trypsin solution at different pH values. The student's results are shown in the table.

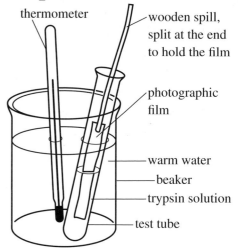

pH	2	4	6	8	10
Time (min)	> 10	7.5	3.6	1.2	8.3
Rate/min	O	0.13			

Remember, rate = 1/time

Guided

(a) Complete the table by calculating the rate of reaction at each pH. **(2 marks)**

(b) Draw a suitable graph to show the effect of pH on the rate of reaction.

A suitable graph would be rate of reaction against pH. Make sure you include a title and label the axes!

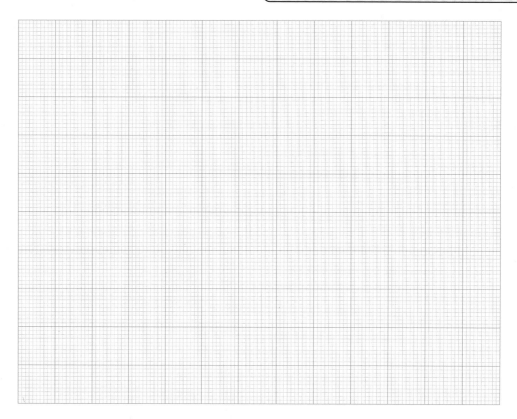

(4 marks)

(c) State **two** ways in which the experiment could be improved.

improvement 1 ....

improvement 2 .... **(2 marks)**

The importance of enzymes

Guided

1 Complete the following table.

Enzyme	Digests	Product(s)
amylase	starch	
lipase		
protease		amino acids

(3 marks)

2 (a) Explain why different digestive enzymes are needed in the digestive system.

....

....

....

.... **(2 marks)**

(b) Explain the importance of enzymes as biological catalysts in building the molecules needed in cells and tissues.

....

....

....

.... **(2 marks)**

3 Biological washing powders contain enzymes that help to break down food stains on clothes.

(a) Eggs are rich in protein. Explain what type of enzyme is needed to remove egg stains from clothes.

....

....

....

.... **(2 marks)**

(b) Explain why biological washing powders work better below 40 °C.

Think about what biological washing powders contain and what effect temperature might have.

....

....

....

....

.... **(3 marks)**

9

🧪 Practical skills Using reagents in food tests

1 (a) (i) State one safety precaution when carrying out food tests.

.... **(1 mark)**

(ii) State one hazard associated with the test for starch.

.... **(1 mark)**

⟩**Guided**⟩ (b) Describe how you would test for the presence of lipids (fats and oils) in a sample of food.

Mix the food with ethanol ....

....

....

....

.... **(3 marks)**

2 A student investigated the changes in broad bean seeds during germination.

(a) In the first experiment, seeds were ground up before germination, then tested for food substances. More seeds were ground up and tested 4 days after germination. The results are shown in the table.

Food test	Observation	
	Before germination	**4 days after germination**
iodine solution	intense blue-black colour	slight blue-black colour
Benedict's solution	blue	red precipitate
biuret test	pale purple	pale purple

Explain what this experiment shows about the changes in food substances during the germination of the broad bean seeds.

> You don't have to understand the detail of germination to answer this question – just focus on the results of the food tests.

....

....

....

.... **(3 marks)**

(b) In the second experiment, the student followed this method:
1. Soak a broad bean seed in water, then cut it in two.
2. Place the cut face on a Petri dish containing starch agar.
3. Leave the plate for 48 hours at 20 °C.
4. Remove the seed, then add iodine solution to the plate. Record any changes seen.

The student noted that the starch agar turned blue-black everywhere, except where it had been covered by the seed. Describe how this result explains the results of the first experiment.

....

....

.... **(2 marks)**

Using calorimetry

1 A student used a calorimeter to test whether the energy values printed on the packaging of a range of foods was correct.

(a) Explain how calorimetry can be used to measure the energy value of food.

....

....

....

.... **(2 marks)**

(b) The student measured the rise in temperature of $20\,cm^3$ of water in the calorimeter when each food was burnt. Some of the student's results are shown in the table.

> Guided

	Brown bread	Biscuit	Dried apricot	Crisps
Energy value given on pack (kJ per 100 g)	1034	1979	1075	2139
Mass of food sample burnt (g)	2.5	2	5	2
Temperature rise of water (°C)	35	24	33	26
Energy value of food sample burnt (J)	2940			
Energy value of food (kJ per 100 g)	118			

(i) It takes 4.2 J of energy to raise the temperature of $1\,cm^3$ water by 1 °C. Complete the table by calculating the energy value in kJ of the food sample burnt and therefore the calculated energy value in kJ per 100 g of each food. Use the space below for working.

> Always show your working – there are marks for correct method even if you get the wrong answer. Also, beware of units – make sure you convert from J to kJ!

(3 marks)

(c) (i) Compare the student's results with the values printed on the packaging.

.... **(1 mark)**

(ii) Give **two** reasons why the student's results might be different.

....

.... **(2 marks)**

Getting in and out of cells

1 Define diffusion.

....

....

....

.... **(2 marks)**

> **Guided**

2 Compare and contrast **diffusion** and **active transport**.

Both ....

....

Active transport requires ....

.... **(2 marks)**

> **Guided**

3 (a) Explain what is meant by the term **osmosis**.

| Make sure you use the terms 'water', 'partially permeable membrane' and 'movement' in your answer. |

Osmosis is the net movement of .... across a ..

....

from a low ....

to a high .... **(4 marks)**

(b) The blood in the lungs contains less oxygen and more carbon dioxide than the air. Explain why oxygen moves from the air to the blood and carbon dioxide moves from the blood to the air.

....

....

....

....

....

.... **(3 marks)**

(c) When starch is digested to glucose it is important that all the glucose is absorbed from the small intestine. Explain why this process requires energy.

....

....

....

.... **(2 marks)**

Practical skills **Osmosis in potatoes**

1 Describe how you would investigate osmosis in potatoes using potato pieces. You are provided with solutions of different sucrose concentrations. You should include at least **two** steps that you should use to ensure the accuracy of your results.

> Guided

Cut pieces of potato, making sure ....

....

....

Remove from the solution, then ....

.... **(4 marks)**

2 The table shows the results of an experiment to investigate osmosis in potatoes.

Sucrose concentration (mol dm^{-3})	Initial mass (g)	Final mass (g)	Change in mass (g)	Percentage change (%)
0.0	19.15	21.60	2.45	12.8
0.1	18.30	19.25	0.95	
0.2	15.32	14.85	−0.47	
0.3	16.30	14.40	−1.90	
0.5	18.25	16.00	−2.25	
1.0	19.50	17.20	−2.30	

> Guided

(a) Complete the table by calculating the percentage change in mass. Use the space below for your working.

> You probably wouldn't have to do as many calculations as this in the exam, but it is good practice!

(2 marks)

(b) (i) Use the axes below to plot a graph of these results.

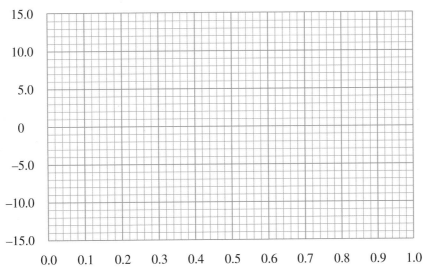

> Remember to label the axes, including the units used.

(3 marks)

(ii) Use your graph to estimate the solute concentration of the potato cells.

solute concentration of potato cells = ... mol dm^{-3} **(1 mark)**

Extended response – Key concepts

The uptake of substances by yeast cells was studied using this method.

1. Add red dye to a suspension of yeast cells. Incubate the cells at 25 °C and observe any changes.

2. Repeat step 1 but incubate the cells at 5 °C.

3. Heat some yeast cells to 60 °C for 2 minutes, then cool them before repeating step 1.

The table shows the results.

Temperature of suspension	Appearance of yeast cells	Appearance of solution
25 °C	red after 30 minutes	colourless after 30 minutes
5 °C	colourless after 30 minutes	still red after 30 minutes
	red after 120 minutes	colourless after 120 minutes
heat-treated at 60 °C then cooled to 25 °C	colourless after 120 minutes	still red after 120 minutes

Explain what this experiment shows about the movement of coloured dye into the yeast cells. Describe other experiments you could do to confirm your conclusion.

You will be more successful in extended response questions if you plan your answer before you start writing. The question asks you to draw conclusions from the data and explain what is happening. Think about the appearance of the solution as well as the appearance of the cells.

Your answer should include the following:

• Describe ways in which the dye could enter the cells – explain the significance of the process being slower at 5 °C than at 25 °C.

• Explain the effect heating would have on the yeast cells. You need to use scientific language, not just 'It killed the cells'.

• Make sure that you include suggestions for further experiments, such as how you might use a microscope to look at what is happening.

Do not forget to use appropriate scientific terminology. Here are some of the words you should include in your answer:

active transport   enzymes   concentration gradient   denatured

---

....

....

....

....

....

....

....

....

....

....

....   **(6 marks)**

Mitosis

1 (a) Read the following statements about mitosis. Which statement is correct? Tick **one** box.

    ☐ **A** A parent cell divides to produce two genetically different diploid daughter cells.

    ☐ **B** A parent cell divides to produce two genetically identical diploid daughter cells.

    ☐ **C** A parent cell divides to produce four genetically identical haploid daughter cells.

    ☐ **D** A parent cell divides to produce four genetically different haploid cells. **(1 mark)**

    (b) List the stages of mitosis in the order they happen, starting with interphase.

    .... **(1 mark)**

2 A cell divides by mitosis every hour. State how many cells there will be after four hours.

> **Guided**

    To start with there is 1 cell; after 1 hour this divides into 2 cells. After 2 hours
    4 cells. After 3 hours 8 cells. After 4 hours .... **(1 mark)**

3 The photograph shows a slide of cells from an onion root tip at different stages in mitosis.

    (a) Name the two stages of mitosis labelled A and B.

    A ....

    B .... **(2 marks)**

    (b) Give a reason for each answer in part (a).

    A ....

    B .... **(2 marks)**

    (c) In filamentous algae, telophase is not followed by cytokinesis. State what the result will be.

> Telophase ends when the nuclear membranes re-form. Think about what normally occurs next and what might be the result if it didn't happen.

    ....

    ....

    ....

    .... **(2 marks)**

Cell growth and differentiation

1 (a) Give the name of a fertilised egg in animals.

.... **(1 mark)**

(b) State the type of cell division that occurs after an egg is fertilised.

.... **(1 mark)**

2 Plant cells divide by mitosis.

(a) State the name of the type of plant tissue where mitosis occurs rapidly.

> Remember that plants grow when cells divide and when cells elongate.

.... **(1 mark)**

(b) Describe how plant cells increase in size following mitosis.

....

.... **(2 marks)**

3 (a) Complete the table to show whether the different specialised cells are animal or plant cells.

> Guided

Type of specialised cell	Animal or plant
sperm	animal
xylem	
ciliated cell	
root hair cell	
egg cell	

(3 marks)

(b) Give the name of one other type of specialised cell found in **plants** and one in **animals**.

plants ....

animals .... **(2 marks)**

4 Growth in animals happens over a particular period of the animal's lifespan. Growth happens through cell division and when cells in the animal differentiate.

(a) Explain what is meant by the term **differentiate**.

....

....

....

.... **(2 marks)**

(b) Explain why cell differentiation is important in animals.

....

....

....

.... **(2 marks)**

Growth and percentile charts

1   A midwife will measure the growth of a baby in different ways. The graph shows some percentile charts for the head circumference measurement for young children.

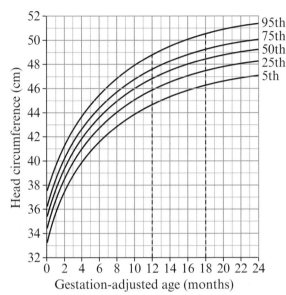

> Graphs like this sometimes look complicated – but remember that the curves are all labelled, so you can see what each one refers to.

> Note that dashed guidelines have been put in to help you answer this part of this question. These help to show how you get the reading for both measurements from the graph – you can then subtract one number from the other to get the final answer.

(a)   The median head circumference is described by the line where half the babies have a greater circumference, and half have the same or a smaller circumference. Which percentile curve shows the median rate of growth for babies?

☐ **A** 5th percentile   ☐ **B** 25th percentile   ☐ **C** 50th percentile   ☐ **D** 75th percentile

(1 mark)

(b)   Use the graph to calculate the change in head circumference for a baby that lies on the 25th percentile curve between 12 and 18 months old. Show your working.

change in circumference .... cm   **(2 marks)**

2   Growth in seedlings can be investigated by measuring the mass of seedlings of different ages.

(a)   One seedling increased in mass from 12.75 g to 15.35 g over a period of 7 days. Calculate the percentage increase in mass for this seedling. Show your working.

Guided

15.35 – 12.75 = ...g

(.............../12.75) × 100 = ... %   **(2 marks)**

(b)   Describe **one** other way you could measure the growth of the seedlings.

....

....

....

....   **(2 marks)**

Stem cells

Guided

1 (a) In animals, stem cells are found in both adults and embryos.

Describe **two** ways in which adult and embryonic stem cells are different from each other.

> When you are asked to describe differences, remember that for each difference you have to say something about both the things you are comparing.

All the cells in an embryo are ...., but in an adult, stem cells are

found only ....

Embryonic stem cells can differentiate into...

but adult stem cells can only differentiate into .... **(2 marks)**

(b) (i) Give the name of the tissue where plant stem cells are found.

.... **(1 mark)**

(ii) Name **two** places in a plant where you would find stem cells.

.... and .... **(2 marks)**

2 (a) Describe **one** function of adult stem cells.

....

.... **(1 mark)**

(b) Describe **one** difference between an embryonic stem cell and a differentiated cell.

....

.... **(1 mark)**

> To answer the following questions, think about what happens in a tissue transplant as well as what the different types of stem cell are capable of.

3 Parkinson's disease is caused by the death of some types of nerve cells in the brain.

(a) Describe how embryonic stem cells could be used to treat Parkinson's disease.

....

.... **(2 marks)**

(b) Another treatment method involves taking the patient's own cells (e.g. skin cells) and turning them into a type of stem cell called IPSCs. Give **one** advantage and **one** disadvantage of each method.

(i) embryonic cells

advantage: ....

disadvantage: .... **(2 marks)**

(ii) IPSCs

advantage: ....

disadvantage: .... **(2 marks)**

The brain and spinal cord

1 (a) Name the two parts of the central nervous system.

.... **(1 mark)**

(b) Describe **one** function of each of the following parts of the brain:

(i) medulla oblongata

.... **(1 mark)**

(ii) cerebellum

.... **(1 mark)**

(c) Describe **two** functions of the cerebral hemispheres.

....

.... **(2 marks)**

2 Sarah goes for a run whilst listening to her MP3 player. During the run she recognises a friend across the road and waves to her.

> **Guided**

Describe the functions of the different parts of Sarah's brain during her run.

Her running is coordinated by the ...., which controls

.... and keeps her balanced.

The .... interpret the sensory information from her

ears while listening to music and also from her eyes when she sees her friend.

Her heart rate and breathing rate are controlled by the ....

Waving to her friend is controlled by the .... **(4 marks)**

3 A man was involved in a fall and fractured his spine. Explain why he became permanently paralysed from the waist down.

> Think about how much was paralysed and what this suggests about where the damage to the spine might be. The question also says 'permanently paralysed' so make sure you address that in your answer as well.

....

....

....

.... **(4 marks)**

Treating damage and disease in the nervous system

1 Describe the function of the following:

(a) the skull

.... **(1 mark)**

(b) the spine

.... **(1 mark)**

2 (a) Explain why some medicines that are effective elsewhere in the body are not effective in treating diseases in the central nervous system.

....

....

.... **(2 marks)**

(b) Benign brain tumours are very slow growing and do not spread throughout the body, but they can still be life-threatening.

> Just because a 'benign' tumour is not cancerous does not mean it is harmless. Think about what damage it might cause as well as the risks of surgery.

(i) Explain why it is difficult to treat brain tumours.

....

....

.... **(2 marks)**

(ii) Why might a surgeon recommend removal of a benign tumour, even though it was not cancerous?

....

.... **(1 mark)**

(iii) Explain how a person who had an operation to remove a benign brain tumour could be left permanently paralysed.

....

.... **(2 marks)**

3 Brain tumours are often treated with radiotherapy to reduce the size of the tumour and then with surgery to remove the tumour.

> Guided

(a) Explain the reason for reducing the size of the tumour before surgery.

To make it easier to remove .... without

removing .... **(2 marks)**

(b) Explain why radiotherapy is sometimes used after surgery to remove a brain tumour.

....

.... **(2 marks)**

Neurones

1 State the function of each of the following.

> Don't say that neurones 'carry messages'; you have to be more specific and talk about electrical impulses.

   (a) dendron.. **(1 mark)**

   (b) axon.. **(1 mark)**

2 The diagram shows a sensory neurone.

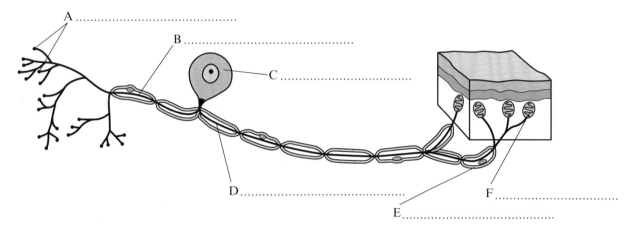

   Label the parts A – F of the sensory neurone. Write your answers on the diagram.   **(3 marks)**

3 (a) Describe one way that the structure of a sensory neurone differs from the structure of a motor neurone.

   The cell body of a sensory neurone is..

   ....

   The cell body of a motor neurone is ....

   .... **(2 marks)**

   (b) Explain how the structure of a motor neurone is related to its function.

   ....

   ....

   .... **(3 marks)**

4 The table shows the speed at which nerve impulses are carried along two types of neurone.

   (a) Explain why the speed of transmission is different in the two types of neurone.

Type of neurone	Speed of transmission (m/s)
myelinated	25
unmyelinated	3

   ....

   .... **(2 marks)**

   (b) In multiple sclerosis (MS), the myelin sheath surrounding motor neurones is destroyed. Explain what effect this would have on the movement of a person with MS.

   ....

   .... **(2 marks)**

Responding to stimuli

1 The diagram shows a junction where neurone X meets neurone Y.

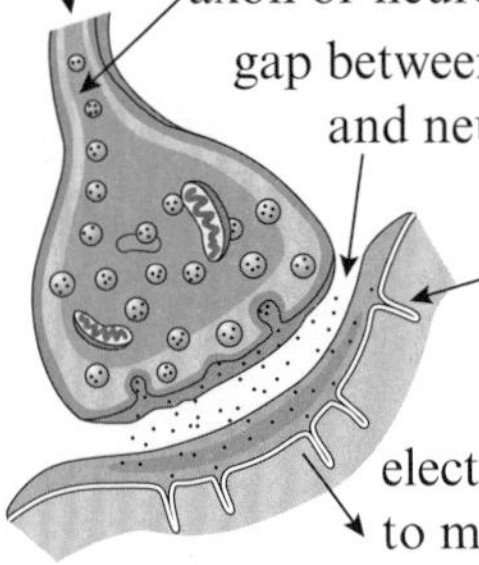

electrical impulse
axon of neurone X
gap between neurone X and neurone Y
neurone Y
electrical impulses to muscle

(a) State the name given to the junction between two neurones.

.... (1 mark)

(b) Explain which neurone (X or Y) on the diagram is a motor neurone.

....

.... (2 marks)

Guided

(c) Describe how neurones X and Y communicate.

When an electrical impulse reaches the end of neurone X it causes the release

of .... into the gap between the neurones. This substance ....

.... across the .... and causes neurone Y to ....

.... (4 marks)

2 The diagram shows a reflex arc.

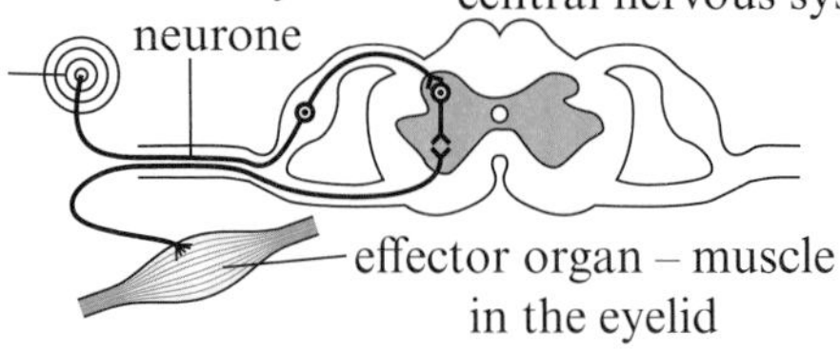

(a) Describe the pathway taken by the nerve impulse in this reflex arc.

stimulus
sensory neurone
central nervous system
effector organ – muscle in the eyelid

....

....

....

....

.... (3 marks)

(b) What is the stimulus in this reflex arc? Give a reason for your answer.

> 'Give a reason' means you have to say something that supports your answer.

....

.... (2 marks)

3 Explain the survival advantage of reflex responses.

....

....

....

....

.... (3 marks)

The eye

1

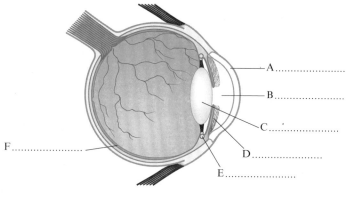

A.....................

B.....................

C.....................

D.....................

E.....................

F.....................

(a) Label the parts A – F on the diagram of the eye...(3 marks)

(b) Explain how the structures of parts A and C are related to their function.

....

....

....

.... (3 marks)

Guided

2 Explain how the iris is adapted to its function.

The iris changes its size by ....

....

It does this to control .... (2 marks)

Guided

3 (a) Light from a distant object falls on the eye.

Describe what happens to the light rays when they enter the eye so that they form a sharp image.

To form an image the light rays must .... onto the ....

This occurs as ....

.... (3 marks)

(b) Describe what happens in the eye to allow it to focus on an object that is close.

....

.... (2 marks)

4 Explain why animals that hunt at night have mostly rod cells in their retinas.

> The question only mentions rod cells, but you will need to talk about cone cells as well in your answer.

....

....

.... (3 marks)

Eye problems

1   The diagram shows a lens in spectacles correcting a sight problem and producing a sharp image on the retina.

(a) State the sight problem being corrected in the diagram.

.... **(1 mark)**

Guided

(b) Explain why the sight correction shown in the diagram is successful.

The lens makes the light rays ....so that the image moves

.... **(2 marks)**

(c) (i)   State a precaution that must be taken if using contact lenses instead of spectacles to correct sight problems.

.... **(1 mark)**

(ii) Contact lenses allow oxygen to pass through them. Give a reason why this is necessary.

> Remember that the cornea is made up of living cells.

.... **(1 mark)**

2   Explain why a person with a cataract has blurred vision.

....

....

.... **(2 marks)**

3   Explain why many people need reading glasses as they get older.

Guided

As they get older the lens does not..

and so they cannot focus on..

.... **(2 marks)**

4   (a) State the type of cell in the retina that detects colour.

.... **(1 mark)**

(b) Describe the cause of red-green colour blindness.

....

....

.... **(2 marks)**

Extended response – Cells and control

Fertilisation of a human egg cell produces a zygote, a single cell that eventually gives rise to every different type of cell in an adult human.

Describe the role of mitosis in the growth and development of a zygote into an adult human.

You will be more successful in extended response questions if you plan your answer before you start writing. Take care, because the question mentions fertilisation but it is really about growth and specialisation. Do not be tempted to talk about sexual reproduction – that is in the next topic.

Your answer should include the following:

• mitosis and cell division causing growth (from embryo to adult), and its importance in repair and replacement of cells

• cell differentiation to produce specialised cells

• the role of stem cells in the embryo as well as in the adult.

Do not forget to use appropriate scientific terminology. Here are some of the words you should include in your answer:

cell cycle   replication   diploid   daughter cells   specialise   differentiate

---

....

....

....

....

....

....

....

....

....

....

....

....

....

.... **(6 marks)**

Asexual and sexual reproduction

1   Complete the table to compare features of sexual and asexual forms of reproduction.

Feature	Sexual reproduction	Asexual reproduction
need to find a mate		
mixing of genetic information	mixes genetic information from each parent	no mixing of genetic information
characteristics of offspring		

(3 marks)

2   During the growing season, strawberry plants send out runners. Where a runner touches the ground a new plant develops. Later in the summer the original plant produces flowers that are fertilised and produce fruits with seeds. Animals eat the fruits and deposit them in their faeces far from the original plant.

(a) Explain which of these is sexual reproduction and which is asexual reproduction.

> 'Explain' means you must identify each type of reproduction **and** give a reason.

runners ....

....

fruits ....

.... **(4 marks)**

(b) Give **one** benefit to the strawberry plant of each type of reproduction.

> Make sure that you give a benefit and say why it is a benefit for each type of reproduction.

....

....

.... **(2 marks)**

(c) Give **one** disadvantage to the strawberry plant of each type of reproduction.

....

....

.... **(2 marks)**

3   Describe **two** ways in which sexual reproduction requires organisms to expend more energy compared to asexual reproduction.

....

....

.... **(2 marks)**

Meiosis

1 Human gametes are haploid cells. During sexual reproduction, the gametes fuse to produce a zygote.

   (a) Describe what is meant by:

      (i) haploid

      .... **(1 mark)**

      (ii) gametes

      .... **(1 mark)**

   (b) State the name of the male sex cells and the female sex cells in humans.

   male ....

   female .... **(2 marks)**

2 A cell contains 20 chromosomes. It divides by meiosis.

   (a) State the number of chromosomes in each daughter cell.

   .... **(1 mark)**

   (b) Explain why the daughter cells are not genetically identical.

   ....

   .... **(2 marks)**

3 The diagram below shows a cell with two pairs of chromosomes undergoing meiosis.

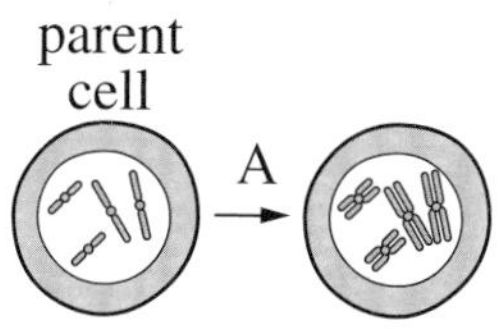

   parent
   cell

   (a) State the name of the process indicated by letter **A** in the diagram.

   .... **(1 mark)**

   (b) Complete the diagram above to show how daughter cells are formed.   **(3 marks)**

   > Use the drawing as a guide. Make sure you draw the chromosomes as they are shown, paying attention to the relative sizes.

4 Describe the importance of the two types of cell division, mitosis and meiosis.

> Guided

   Mitosis maintains the .... and produces cells that are

   ....to the parent cell. It is used for ....

   Meiosis creates .... that have .... the

   number of .... Fertilisation restores the

   .... **(5 marks)**

DNA

1   Our chromosomes contain genetic information. This information is held in our DNA.

   (a) State the name used to describe all the DNA of an organism.

   .... **(1 mark)**

   (b) Describe the difference between chromosomes, genes and DNA.

   > This question is best answered by thinking of the definition of each of these terms.

   A chromosome consists of a long molecule of ....

   ....

   .... **(3 marks)**

2   (a) What name is given to the shape of a DNA molecule?

   .... **(1 mark)**

   (b) The DNA molecule is made up of a series of bases.

      (i)   State the number of different bases present in DNA.

      .... **(1 mark)**

      (ii) Describe how the two strands of the DNA molecule are linked together.

      ....

      .... **(1 mark)**

3   The diagram shows a section of DNA.

   (a) DNA is a polymer. Give **one** piece of evidence from the diagram that DNA is a polymer.

   ....

   ....
   (1 mark)

   (b) Identify the components A, B and C of the DNA structure. Write your answers on the diagram.   **(3 marks)**

   A...........................

   B...........................

   C...........................

   > You will not be expected to draw this structure from memory, but you may be expected to label the parts shown.

4   The sequence of bases on one strand of DNA was ATGGGC.

   (a) Give the order of the complementary bases on the other strand.

   .... **(1 mark)**

   (b) Explain the order that you have written.

   ....

   .... **(2 marks)**

Protein synthesis

1 Describe how the process of transcription produces a strand of mRNA from DNA.

> **Guided**

RNA polymerase binds to a non-coding ....

and then moves along the .... strand adding ....

....

....

....

.... **(4 marks)**

2 The process of translation occurs in the cytoplasm.

   (a) State exactly where in the cytoplasm translation occurs.

.... **(1 mark)**

   (b) State how mRNA enters the cytoplasm.

.... **(1 mark)**

3 The myoglobin protein consists of a single polypeptide of 153 amino acids.

   (a) (i) Calculate the number of codons needed to produce the myoglobin polypeptide.

.... **(1 mark)**

     (ii) Calculate the number of bases in the mRNA for myoglobin.

.... **(1 mark)**

   (b) Describe how the myoglobin protein is formed by the process of translation.

> This question is about translation, so don't talk about transcription in your answer. Make sure you cover all the steps needed to make the final myoglobin protein.

....

....

....

....

....

....

....

.... **(4 marks)**

Gregor Mendel

1   Before the work of Mendel, scientists thought that 'blending' caused variation in inherited characteristics.

(a) Describe how characteristics such as red hair could not be explained by 'blending' of the parents' characteristics.

....

.... **(2 marks)**

> **Guided**

(b) In his work with pea plants, Mendel chose to work with characteristics such as pea shape (round or wrinkled), pea colour (yellow or green) or plant height (tall or short). Explain the importance of his choice of characteristics.

These could not be caused by .... because ....

.... **(2 marks)**

2   Mendel carried out experiments with pea plants. In Experiment 1, he started with pure-bred plants that produced yellow seeds. He crossed these plants with pure-bred plants that produced green seeds. In Experiment 2, he crossed pure-bred plants that produced round seeds with pure-bred plants that produced wrinkled seeds. In both experiments, he crossed plants from the first generation with each other to produce a second generation. The table shows his results.

	Experiment 1		Experiment 2	
Parents	yellow seeds	green seeds	round seeds	wrinkled seeds
First generation	all yellow seeds		all round seeds	
Second generation	349 yellow seeds	112 green seeds	595 round seeds	193 wrinkled seeds

> In each of the following questions you need to **explain**, so make sure you give some evidence to support your answer.

(a) Explain why it was important that Mendel used pure-bred seeds.

....

.... **(2 marks)**

(b) Explain why Mendel pollinated each flower by hand and then placed a bag over the flower.

....

.... **(2 marks)**

(c) Explain what Mendel concluded from these two experiments.

....

....

.... **(3 marks)**

(d) In a third experiment, Mendel took pure-bred plants with yellow round seeds and crossed them with pure-bred plants with green wrinkled seeds. Predict what type of seeds would have been produced in the first generation.

.... **(1 mark)**

Genetic terms

1. Eye colour in humans can be controlled by two alleles of the eye colour gene. One recessive allele (b) codes for blue and one dominant allele (B) codes for brown.

> You need to know what recessive, genotype, phenotype, homozygous and heterozygous mean.

(a) (i) State what is meant by alleles.

.... **(1 mark)**

(ii) Using eye colour as an example, explain the difference between the terms **genotype** and **phenotype**.

....

....

....

.... **(2 marks)**

(b) State the following genotypes for eye colour:

homozygous blue: ....

homozygous brown: ....

heterozygous: .... **(3 marks)**

(c) A girl has blue eyes. Explain what her genotype must be.

....

....

....

.... **(2 marks)**

2. Mendel used the results from his experiments to devise his three laws of inheritance.
   1. Each gamete receives only one factor for a characteristic.
   2. The version of a factor that a gamete receives is random and does not depend on the other factors in the gamete.
   3. Some versions of a factor are more powerful than others and always have an effect in the offspring.

   Mendel did not know what these 'factors' actually were.

   Explain how our understanding of genes and chromosomes has confirmed his laws.

> **Guided**

There are two copies of each chromosome in body cells ....

....

....

....

.... **(4 marks)**

Monohybrid inheritance

1 Two plants both have the genotype Tt. The two plants are bred together.

The allele that makes the plants grow tall is represented by T, and the allele that makes plants shorter is represented by t.

> Percentage probabilities from Punnett squares will always be 0, 25%, 50%, 75% or 100%, depending on the number of squares with a particular genotype (0, 1, 2, 3 or 4 squares). In fractions, probabilities will always be 0, $\frac{1}{4}$, $\frac{1}{2}$, $\frac{3}{4}$, or 1.

(a) Complete the Punnett square to give the gametes of the parents and the genotypes of the offspring.

gametes of parent 1

gametes of parent 2

> Take great care to complete the square correctly and use the right letters.

(2 marks)

Guided

(b) State and explain the percentage of the offspring from this cross that will be short.

25% of the offspring from this cross will be short. I know this because

....

.... **(2 marks)**

(c) Determine the probability of the offspring from this cross being tall. Express your answer as a fraction.

....

.... **(1 mark)**

Guided

2 Fur colour in mice can be represented by two alleles, G and g. Two parent mice were bred, and produced a total of 40 offspring. 50% of the offspring were white, which is the recessive characteristic and the rest were grey.

Complete the genetic diagram to show this cross and show the genotypes of the parents.

> With this question it might be easier to start with what you know – the phenotypes of the offspring – and then work backwards.

Parent genotypes	....		....	
Gametes	....	....	....	....
Genotype of offspring	....	....	....	....
Phenotype of offspring	grey	white	grey	white ....

(4 marks)

Family pedigrees

1 Two healthy parents have a child who has sickle-cell anaemia, a condition caused by a recessive allele. Which **one** of the following is true? **(1 mark)**

> Questions like this can be tricky! Some answers might be true in general, but not in this particular case. You need to pick the one that is true **and** applies to this example.

☐ **A** Both parents are homozygous for the sickle-cell allele.

☐ **B** One parent is homozygous for the sickle-cell allele and the other is homozygous for the normal allele.

☐ **C** Both parents are heterozygous for the sickle-cell allele.

☐ **D** One parent is heterozygous for the sickle-cell allele and the other is homozygous for the normal allele.

2 This family pedigree shows the inheritance of cystic fibrosis (CF).

CF is a genetic condition in humans caused by a recessive allele.

(a) State how many cystic fibrosis alleles an individual must inherit in order to show the symptoms of CF.

....
(1 mark)

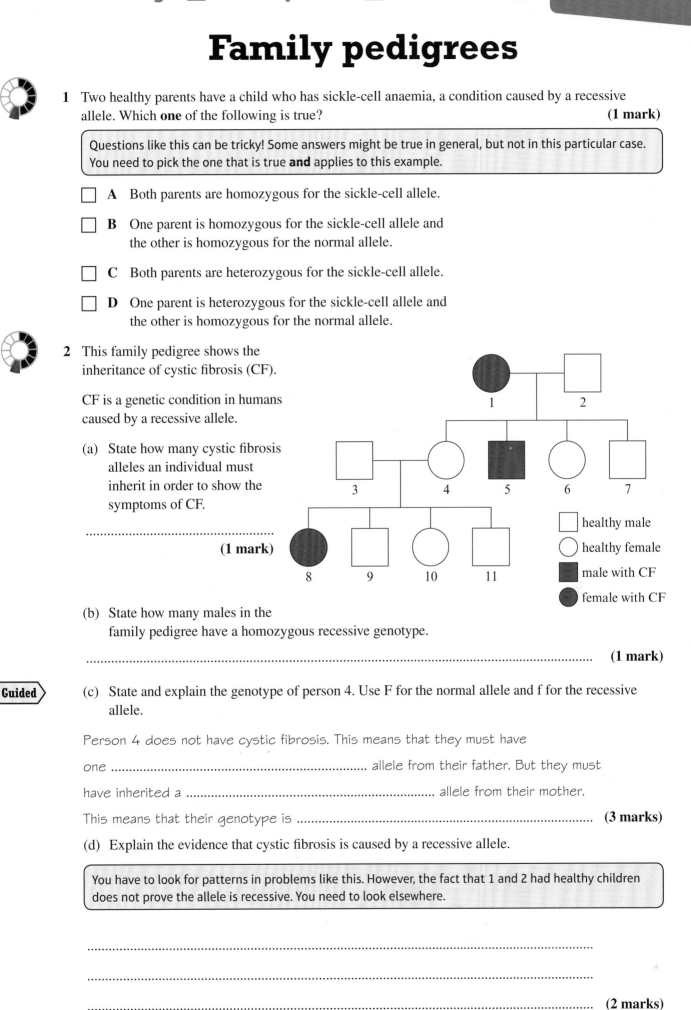

healthy male

healthy female

male with CF

female with CF

(b) State how many males in the family pedigree have a homozygous recessive genotype.

.... **(1 mark)**

Guided

(c) State and explain the genotype of person 4. Use F for the normal allele and f for the recessive allele.

Person 4 does not have cystic fibrosis. This means that they must have

one .... allele from their father. But they must

have inherited a .... allele from their mother.

This means that their genotype is .... **(3 marks)**

(d) Explain the evidence that cystic fibrosis is caused by a recessive allele.

> You have to look for patterns in problems like this. However, the fact that 1 and 2 had healthy children does not prove the allele is recessive. You need to look elsewhere.

....

....

.... **(2 marks)**

33

Sex determination

1 (a) A baby girl is born. Explain which sex chromosome was in the sperm that fertilised the egg.

....

.... **(2 marks)**

(b) (i) Complete the Punnett square to show the sex chromosomes of both parents and all possible children.

> This is a Punnett square but you could also use a genetic diagram to show how X and Y chromosomes combine.

Father

X

Mother { X

(2 marks)

(ii) State the sex of the child in the shaded box.

.... **(1 mark)**

2 (a) A couple who have a girl wish to have a second child. Explain the chance of the couple's second child being a boy.

....

....

....

....

....

.... **(3 marks)**

(b) Read this statement:

> If a couple have had children and they are all girls, then the next child is more likely to be a boy.

Discuss whether you think this statement is correct.

....

....

....

.... **(2 marks)**

Inherited characteristics

1   Using blood groups as an example, describe what is meant by codominance.

.... **(1 mark)**

2   A couple had four children, half of whom had group A blood and half of whom had group AB blood. The man was group A.

(a)   State the genotype of the children.

group A blood: ....

group AB blood: .... **(2 marks)**

(b)   State and explain the genotype and blood group of the mother.

> A Punnett square might help to solve this one, but is not essential.

....

....

.... **(3 marks)**

(c)   The mother had a fifth child who had group B blood. Explain why the father of her first four children could not be the father of this child.

> **Guided**

A child who is group B must receive .... from each parent,

.... **(2 marks)**

3   A couple are both blood group AB.

Use a Punnett square to work out the possible blood groups of their children and the probability of each.

> Don't forget that you can express probability as a percentage or a fraction.

Probability of each blood group: .... **(3 marks)**

Variation and mutation

Guided

1 What are the causes of differences between the following?

(a) the masses of students in a year 7 class

Students in a year 7 class will show differences in mass caused by ....

variation as well as .... variation. **(2 marks)**

(b) a pair of identical twins

Identical twins will only show differences caused by .... variation. **(1 mark)**

2 Mr and Mrs Davies have six children. The table shows the heights of each of the six children when they reached adulthood.

Child	George	Arthur	Stanley	James	Josh	Peter
Adult height in cm	181	184	178	190	193	179

a) Calculate the mean height of the six Davies children. Show your working out. Give your answer to 1 decimal place.

mean height = ... cm **(2 marks)**

(b) Mr Davies is 192 cm tall and Mrs Davies is 165 cm tall. Mr Davies wonders why his children show a range of different heights. Mrs Davies wonders why the mean height of the children is not the same as the mean of her height and her husband's height. Suggest an explanation that will answer their questions.

> Don't forget to cover both genetic and at least one environmental factor. Make sure you use scientific language such as alleles and inheritance in your answer.

....

....

....

....

.... **(4 marks)**

3 (a) State the possible effects of a mutation on the phenotype of an organism.

....

....

.... **(2 marks)**

(b) Cystic fibrosis is caused by a mutation that produces an inactive protein in the lung. Explain how the cystic fibrosis mutation leads to production of an inactive protein.

....

....

....

.... **(3 marks)**

The Human Genome Project

1 (a) State what is meant by the human genome.

....

.... **(1 mark)**

(b) State **two** advantages and **two** disadvantages of decoding the human genome.

advantage 1

A person at risk from a genetic condition will be ....

advantage 2

....

disadvantage 1

....

disadvantage 2

.... **(4 marks)**

2 The *BRCA1* mutation increases a woman's risk of developing breast cancer. Discuss the advantages and disadvantages to a woman of knowing that she has the *BRCA1* mutation.

> **Discuss** means you need to identify the issues being assessed by the question. You need to explore the different aspects of the issue. In this case, these are the advantages and disadvantages involved.

....

....

....

....

....

....

....

.... **(4 marks)**

Extended response – Genetics

The protein p53 is 393 amino acids long and helps to control cell division. People with a mutation of the p53 gene are more likely to develop cancer. With reference to the production of the p53 protein, discuss how the Human Genome Project can help to identify those people who are at risk from cancer linked to mutant p53.

You will be more successful in extended response questions if you plan your answer before you start writing.

You should already know that the two processes involved in making p53 are transcription and translation, so you just need to fill in the details. Then you need to link this to how the Human Genome Project helps to identify people at greater risk of developing cancer.

It is easy to get the processes involved here confused, so be very careful when deciding how to structure your answer. Your answer could explain the following points:

• which part of the DNA contains the information to make p53

• how the information from the DNA is transferred around the cell

• where in the cell the manufacture of p53 takes place

• how the information is used to decide which amino acids make up the final protein

• how each amino acid is added to the chain to make a protein

• whether the mutation will affect the final p53 protein.

....

....

....

....

....

....

....

....

....

....

....

....

....

.... **(6 marks)**

Evolution

1 (a) Describe the work of Darwin and Wallace in the development of the theory of evolution.

....

....

.... **(2 marks)**

(b) Describe the impact of Darwin and Wallace's ideas on modern biology.

....

....

.... **(2 marks)**

2 Explain why, when an environment changes, some organisms within a species survive whereas others die.

> You should use scientific terms such as variation and survival in your answer.

....

....

.... **(2 marks)**

3 When a new species is discovered, a scientist may take some of its DNA to analyse. Explain how this would help establish if this is a new species.

⟩ **Guided** ⟩

It will help .... the new species and to find out

which other ....

.... **(2 marks)**

4 It is important to complete a course of antibiotics.

(a) Explain how stopping a course of antibiotics early can cause antibiotic resistance in bacteria.

> Darwin's theory was about natural selection and the survival of the fittest, so you should relate these to antibiotic resistance in bacteria.

....

....

....

....

.... **(4 marks)**

(b) Explain how this provides evidence for Darwin's theory of evolution.

....

....

....

.... **(3 marks)**

Human evolution

1 Apart from the differences in body hair, using the diagrams of Ardi and Lucy, state three differences between them.

1. ....

....

2. ....

....

3. ....

....

Ardi   Lucy

(3 marks)

2 Some evidence for human evolution has come from the fossil record of the skull. The table below shows some of this evidence.

> You do not need to remember details such as brain sizes but you do need to remember the names and the general trends.

Name of species	Year before present when species first appeared (millions of years ago)	Brain volume (cm³)
Ardipithecus ramidus (Ardi)	4.4	350
Australopithecus afarensis (Lucy)	3.2	400
Homo habilis	2.4	550
Homo erectus	1.8	850

(a) Describe the relationship between when each species first appeared and brain volume.

....

.... **(2 marks)**

(b) The first stone tools are dated from about 2.4 million years ago. Using the table, deduce what may have enabled the use of stone tools.

⟩ **Guided** ⟩   An increase in ....

.... **(2 marks)**

3 The diagram shows two images of stone tools.

(a) Explain how scientists work out the ages of stone tools.

....

A   B

....

....

.... **(2 marks)**

(b) Using the diagram, explain how stone tool A was held. Give reasons for your answer.

....

....

.... **(3 marks)**

Classification

1 Describe the similarities between a human arm and a bat's wing that suggest humans and bats share a common ancestor.

....

.... **(2 marks)**

2 Give **two** reasons why animals and plants are placed in separate kingdoms.

Guided

Plants .... but animals...

Plant cells have ....

but... **(2 marks)**

3 The table shows how some organisms are classified.

Classification group	Humans	Wolf	Panther
kingdom	Animalia	Animalia	Animalia
phylum	Chordata	Chordata	Chordata
class	Mammalia	Mammalia	Mammalia
order	Primate	Carnivora	Carnivora
family	Hominidae	Canidae	Felidae
genus	Homo	Canis	Panthera
species	Sapiens	Lupus	Pardus
binomial name	*Homo sapiens*	*Canis lupus*	*Panthera pardus*

Explain which two organisms in the table are most closely related.

....

....

.... **(2 marks)**

4 Carl Woese proposed that all organisms should be divided into three domains. Complete the table to give the missing information.

Follow the example of the table and comment on the nucleus and genes.

Domain	Distinguishing characteristic of the domain
Archaea	
Eubacteria	
	cells with a nucleus, unused sections in genes

(3 marks)

Selective breeding

1 (a) Describe what is meant by selective breeding.

....

.... **(2 marks)**

(b) Explain how pig breeders could use selective breeding to produce lean pigs with less body fat.

> The principles of selective breeding are the same, even if you aren't familiar with this example.

....

....

....

....

.... **(3 marks)**

2 Food production can be increased by conventional plant breeding programmes.

(a) Explain **three** different characteristics that could be selected for in a crop suitable for use in any country.

....

....

....

....

.... **(3 marks)**

(b) Explain **two** other characteristics that might be selected for in a crop to be grown in a hot, dry part of Africa.

....

....

.... **(2 marks)**

(c) Give a reason why wheat in the United Kingdom has been selected to have a short stem length.

....

.... **(1 mark)**

3 Give **three** risks of selective breeding.

1. Alleles that might be useful in the future ....

2. ....

....

3. ....

.... **(3 marks)**

Genetic engineering

Guided

1 Scientists have produced mice that glow green in blue light. These 'glow mice' contain a gene naturally found in jellyfish. Explain why a glow mouse is described as a genetically modified organism.

Mice do not normally glow, but glow mice have a ....

....

.... **(2 marks)**

2 Scientists may genetically modify crop plants.

(a) Name one crop plant that has been genetically modified.

.... **(1 mark)**

(b) Describe two ways in which genetically modified crops benefit humans.

....

....

.... **(2 marks)**

(c) Genetic modification can make crop plants resistant to insects by introducing certain bacterial genes.

Explain one disadvantage of doing this.

> Think about how these genetically modified plants may affect other living organisms in their environment.

....

....

....

....

.... **(2 marks)**

3 Human insulin can be produced by genetically modified bacteria. Discuss the advantages and disadvantages of this process.

> When asked to discuss, you need to give both sides of an argument. Here you should give at least one advantage and one disadvantage.

....

....

....

....

....

.... **(4 marks)**

Tissue culture

1 Tissue culture is used to produce clones of a plant.

(a) Describe how plant tissue culture is used to produce clones from a parent plant.

....

....

....

....

.... **(4 marks)**

(b) Describe an advantage of producing many clones of a plant.

....

.... **(2 marks)**

(c) Give a disadvantage of producing plants using tissue culture.

> You need to apply your knowledge to give an answer. Consider how having many identical plants might be a problem.

....

.... **(1 mark)**

2 A pharmaceutical company used cell culture to test a new drug for safety before giving it to humans.

(a) Describe how the company would produce the culture.

> You may need to review stem cells on page 18 of the Revision Guide.

....

....

.... **(2 marks)**

⟩ Guided ⟩

(b) Explain the advantages of using cell culture to test a new drug.

Cell culture will be .... and .... and does not have

any .... issues if the drug damages the cells. Also, human cells could

be used which ....

.... **(4 marks)**

Stages in genetic engineering

1 State the meaning of the following terms.

(a) plasmid

.... **(1 mark)**

(b) vector

.... **(1 mark)**

(c) sticky ends

.... **(1 mark)**

2 People with diabetes rely upon insulin. Human insulin can be produced by genetically modified bacteria, produced through genetic engineering. Explain the role of the following in the production of human insulin:

> **Guided**

(a) a human gene

The human gene needed is ....

It is needed because .... **(2 marks)**

(b) enzymes

....

.... **(2 marks)**

(c) bacteria

....

.... **(2 marks)**

3 (a) Explain why the same restriction enzyme is used to cut DNA from a human cell and to cut bacterial plasmids open.

> Remember that restriction enzymes produce DNA fragments with 'sticky ends'.

....

....

....

....

....

.... **(3 marks)**

(b) Describe the role of DNA ligase in genetic engineering.

....

....

....

.... **(2 marks)**

Insect-resistant plants

1 Bt plants are an example of insect-resistant plants. Bt plants have had a gene from the *Bacillus thuringiensis* bacterium introduced into them.

(a) Explain why the gene for Bt toxin has been used to genetically modify some crop plants.

Guided

| Toxins are poisonous substances. |

The Bt toxin is a substance that...

....

....

.... **(3 marks)**

(b) Describe how bacteria are used to produce plants that produce the Bt toxin.

....

....

....

....

.... **(3 marks)**

2 There are advantages and disadvantages to growing genetically modified plants such as the insect-resistant plant containing the Bt gene.

(a) Explain **one** advantage.

....

....

....

.... **(2 marks)**

(b) Some wild plant species are closely related to the genetically modified plants. The Bt gene might transfer to the genomes of these wild plants.

(i) Explain how this transfer could happen.

....

....

....

.... **(2 marks)**

(ii) Explain why the transfer of the Bt gene could be a problem for other species in the environment.

| You need to understand that new scientific developments have advantages, but also disadvantages. |

....

....

....

.... **(2 marks)**

Meeting population needs

1 State **two** agricultural solutions that attempt to meet the need for food of a growing human population.

(a) ....

(b) .... **(2 marks)**

2 Aphids (greenfly) are insect pests that attack many food crops, including potatoes. Two identical plots of potato plants were chosen and infested with aphids. In one plot ladybirds (natural predators of aphids) were also introduced. The graph shows the number of aphids on the potato plants in the two plots during the 10 days after infestation, either with or without ladybirds.

Graph: y-axis "Aphids per leaf" 0–35; x-axis "Days after initial infestation" 0–10. Legend: without ladybirds (diamond), with ladybirds (square).

> Make sure you quote numerical data from the graph in your answer.

(a) Describe how the population of aphids in the absence of ladybirds changed during the 10 day period.

....

....

.... **(3 marks)**

(b) Describe and explain the differences seen when ladybirds are present on the potato plants.

When ladybirds are present, the number of aphids ....

> Guided

because ....

.... **(3 marks)**

3 Give **two** reasons why some farmers use an integrated approach to pest control, combining biological and chemical control methods.

> Biological controls have disadvantages; you should explain what these are and how chemical pesticides might make up for these.

....

....

.... **(2 marks)**

4 *'As human populations grow and we need to produce more food, the advantages of using artificial fertilisers outweigh the disadvantages.'* Assess this statement.

> Decide whether you agree or disagree with this statement. Give reasons that support your view, and reasons against your view.

....

....

.... **(3 marks)**

Extended response – Genetic modification

Compare and contrast the processes of evolution and selective breeding.

> You will be more successful in extended response questions if you plan your answer before you start writing.
>
> In this question, you need to say how these two processes are similar and how they are different.

....

....

....

....

....

....

....

....

....

....

....

....

.... **(6 marks)**

Health and disease

1   According to the World Health Organization (WHO), good health is a state of 'complete physical, social and mental well-being'. State what is meant by the following terms.

(a) physical well-being

.... **(1 mark)**

(b) mental well-being

.... **(1 mark)**

(c) social well-being

.... **(1 mark)**

2   (a) Complete the table by putting a tick in the appropriate box to show whether the disease is communicable or non-communicable.

> **Guided**

Disease	Communicable	Non-communicable
influenza ('flu')	✔	
lung cancer		
coronary heart disease		
tuberculosis		
Chlamydia (a type of STI)		

(3 marks)

(b) Explain why you identified some diseases as communicable and others as non-communicable.

....

.... **(2 marks)**

3   HIV is a virus that can infect humans. HIV makes it easier for other pathogens to infect the human body. Suggest an explanation for how HIV does this.

> Think about what type of cells are infected by the HIV virus.

....

.... **(2 marks)**

4   (a) Explain how viruses cause disease.

....

....

.... **(3 marks)**

(b) Describe **two** ways in which bacteria make us feel ill.

....

....

.... **(2 marks)**

Common infections

Maths skills

1 The table shows the percentage of 15 to 49 year olds with HIV in some African countries.

African country	% of 15 to 49 year olds with HIV in some African countries			
	2006	**2007**	**2008**	**2009**
Namibia	15.0	14.3	13.7	13.1
South Africa	18.1	18.0	17.9	17.8
Zambia	13.8	13.7	13.6	13.5
Zimbabwe	17.2	16.1	15.1	14.3

(a) Identify the country with the largest decrease in the percentage of HIV between 2008 and 2009. Show your working

> First work out what the decrease was for each country. For example Zambia went from 13.6% to 13.5%. If you are not sure – use your calculator!

country with largest decrease .... **(2 marks)**

(b) The data for each African country follows the same overall trend. Use the data in the table to describe this trend.

....

.... **(2 marks)**

2 (a) What kind of pathogen causes *Chalara* ash dieback?

Guided

☐ **A** ~~a virus~~   ☐ **C** a protist

☐ **B** a bacterium   ☐ **D** a fungus   **(1 mark)**

(b) Describe the effects of the pathogen on the trees.

....

.... **(2 marks)**

3 The table shows several diseases, the type of pathogen that causes them and the symptoms (signs of infection). Complete the table by filling in the gaps.

Disease	Type of pathogen	Signs of infection
cholera		watery faeces
	bacterium	persistent cough – may cough up blood-speckled mucus
malaria		
HIV		mild flu-like symptoms at first

(3 marks)

4 *Helicobacter* is a pathogen that causes stomach ulcers.

(a) State the type of pathogen involved.

.... **(1 mark)**

(b) Describe the symptoms it causes in infected people.

....

.... **(2 marks)**

How pathogens spread

1 Which of these statements about malaria is correct?

☐ **A** Malaria is caused by a mosquito that invades liver cells.

☐ **B** The malaria pathogen is a mosquito.

☐ **C** The malaria pathogen is a protist that is spread by a vector, the mosquito.

☐ **D** The malaria pathogen is a mosquito that is spread by a vector, the protist.   **(1 mark)**

2 Complete the table to show ways in which the spread of certain pathogens can be reduced.

Disease	Pathogen	Ways to reduce or prevent its spread
Ebola haemorrhagic fever		keep infected people isolated; wear full protective clothing while working with infected people or dead bodies
tuberculosis	bacterium	

(2 marks)

3 Cholera is a disease that can spread rapidly in disaster areas when drinking water supplies are damaged. Explain **one** way that its spread could be reduced or prevented.

....

....

....

....   **(2 marks)**

4 (a) Explain why bacterial diseases such as cholera are less common in developed countries.

Think about how these diseases are spread and how developed countries are able to control them.

....

....

....

....   **(2 marks)**

(b) Explain why, during the 2014–15 Ebola outbreak, health workers wore full body protection when handling dead bodies.

> **Guided**

To prevent being infected ....

...........................because Ebola virus is present ....

....

....

....

....   **(2 marks)**

STIs

1 State what is meant by an STI.

....

.... **(1 mark)**

2 *Chlamydia* is a pathogen that causes an STI. Which of these statements is correct?

☐ **A** *Chlamydia* is a virus.

☐ **B** A person infected with *Chlamydia* may not realise they are infected.

☐ **C** *Chlamydia* has a lysogenic cycle.

☐ **D** *Chlamydia* cannot be passed from mother to baby during birth.   **(1 mark)**

3 Complete the table.

> Guided

Mechanism of transmission	Precautions to reduce or prevent STI
unprotected sex with an infected partner	using .... during sexual intercourse
	supplying intravenous drug abusers with sterile needles
infection from blood products	

(3 marks)

4 The HIV virus has a lysogenic cycle.

(a) Describe what happens to the virus during the lysogenic cycle.

....

....

....

.... **(2 marks)**

(b) Explain why it can be many years between being infected with HIV and developing AIDS.

....

....

....

....

.... **(2 marks)**

Human defences

1 (a) Describe the role of the skin in protecting the body from infection.

....

.... **(1 mark)**

(b) Describe one chemical defence against infection from what we eat or drink.

.... **(1 mark)**

(c) (i) Name an enzyme, found in tears, that protects against infection.

.... **(1 mark)**

(ii) Describe how the enzyme named in part (i) protects the eyes against infection.

> You need to name the enzyme and say what it does.

....

.... **(2 marks)**

2 The diagram shows a section of epithelium in a human bronchiole, one of the tubes in the lung.

(a) (i) State the name of the substance labelled A.

.... **(1 mark)**

(ii) Describe the role of substance A in protecting the lungs from infection.

.... **(1 mark)**

(b) (i) State the name of the structure labelled B.

.... **(1 mark)**

(ii) Describe the part played by the type of cell labelled C in protecting the lungs from infection.

> **Guided**

The .... on the surface of these cells move in a wave-like motion

....

....

.... **(3 marks)**

(c) Chemicals in cigarette smoke can paralyse the structures labelled B.

Explain why this increases the risk of smokers suffering from lung infections compared with non-smokers.

....

....

.... **(2 marks)**

The immune system

1. Name the type of blood cell that produces antibodies.

   .... **(1 mark)**

2. Describe how lymphocytes help protect the body by attacking pathogens.

   Pathogens have substances called .... on their surface. White blood

   cells called .... are activated if they have .... that fit these

   substances. These cells then ....

   They produce large amounts of antibodies that .... **(5 marks)**

3. The graph shows the concentration of antibodies in the blood of a young girl. The lines labelled A show the concentration of antibodies effective against the measles virus. The line labelled B shows the concentration of antibodies effective against the chickenpox virus.

   > There is a lot to think about in this question so take it one step at a time.

   (a) At the time shown by arrow 1, there was an outbreak of measles. The girl was exposed to the measles virus for the first time in her life.
   Explain the shape of line A in the five weeks after arrow 1.

   ....

   ....

   ....

   ....

   .... **(4 marks)**

   (b) Five months later (shown by arrow 2) there was an outbreak of measles and chickenpox. The girl was exposed to both viruses.
   Explain the shape of line A in the five weeks after arrow 2.

   ....

   .... **(3 marks)**

   (c) Use lines A and B to help you answer these questions.

   (i) State whether the girl had been exposed to the chickenpox virus in the past. Explain your answer.

   ....

   .... **(2 marks)**

   (ii) In the second outbreak of measles, the girl showed no symptom of measles. Explain why.

   ....

   .... **(2 marks)**

Immunisation

1 (a) Children are given vaccinations against many childhood infections. State what is meant by a vaccine.

Guided

A vaccine contains antigens from ....

.... **(2 marks)**

(b) A vaccine prevents a person from becoming ill from infection with a pathogen. This works even if they are exposed to the pathogen a long time after the vaccination. Explain why.

....

....

....

.... **(3 marks)**

(c) Describe **two** disadvantages of vaccination.

....

.... **(2 marks)**

2 In 1998 a group of doctors suggested there was a connection between the MMR (measles, mumps and rubella) vaccine and autism. This made some parents afraid of having their babies vaccinated.

The graph shows how the percentage of babies in the UK who were given the MMR vaccine changed over the following years.

(a) State which year had the lowest rate of vaccination.

....

(1 mark)

(b) Predict what would happen to the number of children suffering from measles in the period 1998–2004. Justify your answer.

....

....

.... **(2 marks)**

(c) The target immunisation rate for measles is 95%. Explain why it is not necessary for 100% of children to be immunised.

> This question is asking you to talk about herd immunity.

....

....

.... **(2 marks)**

Treating infections

1 (a) Which of the following statements is correct?

☐ **A** An antibiotic is produced in the body to fight infection.

☐ **B** Some antibiotics are becoming resistant to bacteria.

☐ **C** Antibiotics are medicines that kill or slow down growth of bacteria in the body.

☐ **D** Antibodies are medicines that kill or slow down growth of bacteria in the body. **(1 mark)**

(b) Explain why antibiotics can be used to treat bacterial infections in people.

....

....

.... **(2 marks)**

2 Colds are caused by viruses. A man has a very bad cold. He asks a pharmacist if an antibiotic such as penicillin would help to cure his cold.

> **Guided**

State, with a reason, whether the pharmacist would advise the man to take penicillin.

The pharmacist's advice would be ....

The man's cold is due to a virus, so the penicillin ....

.... **(2 marks)**

3 Sinusitis causes a stuffy nose. Some patients with sinusitis were divided into two groups. One group was treated for 14 days with antibiotics while the other group did not receive antibiotics. Each day they were asked if they still had symptoms. The results are shown in the graph.

(a) State what you can deduce about the cause of sinusitis from the data.

> You are only asked for a deduction, not an explanation – although you might need to think about the answer to part (b) before you make your deduction!

.... **(1 mark)**

(b) Discuss whether the data supports the use of antibiotics to treat sinusitis.

> Be sure to refer to data in the graph when answering this question.

....

....

....

.... **(2 marks)**

Aseptic techniques

When answering several of the questions on this page, it is not enough to say 'it kills microorganisms'; you have to say why that is important.

1 Give a reason for each of the following practices when working with cultures of microorganisms.

(a) It is important to wipe the workbench with disinfectant before starting work.

.... **(1 mark)**

(b) It is important not to completely seal the Petri dish.

.... **(1 mark)**

(c) Cultures are incubated at 25 °C in school and college laboratories even though they grow faster at 37 °C.

.... **(1 mark)**

(d) It is important to sterilise an inoculating loop after use.

.... **(1 mark)**

2 It is important that people doing experiments to culture microorganisms follow some safety precautions. For each precaution given, explain why it is important.

> **Guided**

(a) The lid of the Petri dish is only opened enough to inoculate the agar plate.

This will .... microorganisms from the air that are..............

.... **(2 marks)**

(b) The inoculating loop is held in a flame before use.

....

.... **(2 marks)**

(c) The lid of the Petri dish is loosely taped down.

....

.... **(2 marks)**

3 A student is culturing some bacteria. Here are the steps he uses:
- Agar jelly is heated to 80 °C.
- The agar jelly is cooled and a sample of bacteria is added when the jelly is at 21 °C.
- The jelly is put into sterilised Petri dishes and it is then warmed to 25 °C for around 6 hours.

State why each of these steps is important in making the bacterial culture safely and efficiently.

....

....

....

....

....

.... **(3 marks)**

Investigating microbial cultures

1 Paper discs were dipped into different antibiotics. They were placed on to a culture of bacteria, which was growing on nutrient agar in a Petri dish. The dish was then incubated for 3 days. The diagram shows the results.

bacterial growth

antibiotic disc

(a) Explain why there are clear areas around each disc.

....

.... **(2 marks)**

(b) Complete the table by using the diagram.

Antibiotic	Diameter of clear area (mm)	Cross-sectional area (mm²)
1		
2		
3		
4		

(4 marks)

(c) Which antibiotic is the most effective in this experiment? Explain your answer.

> Remember to use evidence from the experiment to justify your explanation.

....

....

.... **(2 marks)**

2 (a) Describe how the method in question 1 could be adapted to investigate the effect of concentration of antibiotic on growth of bacteria.

Guided

Use different concentrations of ....

....

....

.... **(3 marks)**

(b) State **two** ways in which you would make sure the investigation gives good results.

1. ....

....

2. ....

.... **(2 marks)**

New medicines

1 Development of a new medicine involves a series of stages. A new medicine can only move to the next stage if it has been successful in the previous stage.

(a) Complete the table to show the correct order of stages of developing a new drug.

> **Guided**

Stage	Order
Testing in a small number of healthy people	
Discovery of possible new medicine	1
Given widely by doctors to treat patients	
Testing in cells or tissues in the lab	
Testing in a large number of people with the disease the medicine will treat	

(2 marks)

(b) (i) Describe **two** stages of preclinical testing in the development of a new medicine.

....

.... **(2 marks)**

(ii) Describe how development of a new medicine ensures that there are no dangerous side effects in humans.

.... **(1 mark)**

(c) Describe the function of a large clinical trial in developing a new drug.

> For three marks you will have to describe all of the functions; pay attention to the word 'large'.

....

....

.... **(3 marks)**

2 Scientists trialled a new medicine that was developed to lower blood pressure. They took 1000 people with normal blood pressure (group A) and 1000 people with high blood pressure (group B). Each group was divided in half; half the volunteers were given the new medicine and the other half were given a placebo (dummy medicine). At the end of the trial, the scientists measured the number of volunteers in each group who had high blood pressure.

The results are shown in the bar chart.

Number of people with high blood pressure at the end of the trial (y-axis: 0, 50, 100, 150, 200, 250, 300, 350, 400, 450, 500)

x-axis categories: Group A given medicine / Group A given placebo / Group B given medicine / Group B given placebo

(a) Explain why it is important for medicine trials to use large numbers of volunteers.

....

.... **(2 marks)**

(b) Use information from the bar chart to evaluate the effectiveness of this medicine.

....

.... **(2 marks)**

Monoclonal antibodies

1 (a) Name the type of mouse cell that makes antibodies.

.... **(1 mark)**

(b) Explain why these types of cells are not suitable for making large quantities of antibodies.

....

.... **(2 marks)**

2 The diagram below shows a pregnancy test stick.

> **Guided**

region containing monoclonal   handle
antibodies to test for pregnancy hormone

(a) Describe what is meant by monoclonal antibodies.

antibodies of one type produced in large quantities by .... cells **(2 marks)**

(b) Describe how the monoclonal antibodies used in the pregnancy test stick are produced.

....

....

....

....

....

.... **(4 marks)**

(c) State why this stick would not respond to a different hormone.

.... **(1 mark)**

3 (a) Explain how a monoclonal antibody can be used to detect cancer cells in the body.

....

....

.... **(2 marks)**

(b) Medicines used to treat cancer can kill healthy cells as well as cancer cells. Describe how monoclonal antibodies can be used to deliver medicines to kill cancer cells.

> Make sure you include the effect on cancer cells as well as on normal cells in your answer.

....

....

....

....

.... **(4 marks)**

Non-communicable diseases

Guided

1 Explain how an infectious disease is different to a non-communicable disease.

An infectious disease is caused by a .... and is passed from ....

.... A non-communicable disease

is not passed .... **(3 marks)**

2 State **three** factors that can affect a person's risk of developing a non-communicable disease.

1. ....

2. ....

3. .... **(3 marks)**

3 The two graphs show the prevalence of coronary heart disease (CHD) in men and women from different ethnic groups in the West Midlands. Prevalence means the percentage of people in that ethnic group who are diagnosed with the disease.

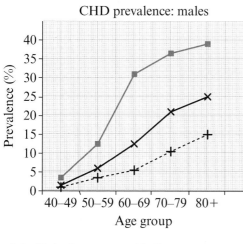
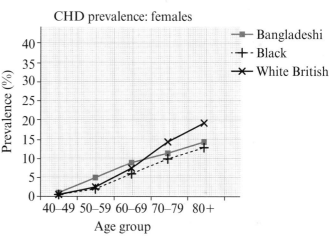

(a) State the group with the:

(i) highest incidence of CHD .... **(1 mark)**

(ii) lowest incidence of CHD.. **(1 mark)**

(b) Discuss the effect of age, sex and ethnic group on the risk of developing CHD. Use the information in the graphs in your answer.

> **Discuss** means you need to identify the issues being assessed by the question. You need to explore the different aspects of the issue. In this case, how the incidence of CHD varies with age, sex and ethnic group.

> Make sure you cover all three factors (age, sex and ethnic group) as well as using data from the graph to support your conclusions.

....

....

....

....

....

.... **(4 marks)**

Alcohol and smoking

1 (a) Explain how alcohol (ethanol) causes liver disease.

....

....

....

....

....

.... **(3 marks)**

(b) State why alcohol-related liver disease is described as a lifestyle disease.

....

.... **(1 mark)**

2 Babies whose mothers smoked while pregnant have low birth weights. Explain why.

....

....

....

.... **(2 marks)**

3 (a) State **two** diseases caused by substances in cigarette smoke.

> The question asks you to state two diseases. Remember that heart attack or stroke are not diseases, they are the result of disease.

....

....

....

.... **(2 marks)**

(b) A stroke is caused by cardiovascular disease in the brain. Explain how smoking can lead to a stroke.

Guided

Substances in cigarette smoke cause blood vessels to ....

....

....

....

....

....

.... **(3 marks)**

Malnutrition and obesity

1 The graph shows the percentage of different age groups with anaemia in a population in the USA during the 1990s.

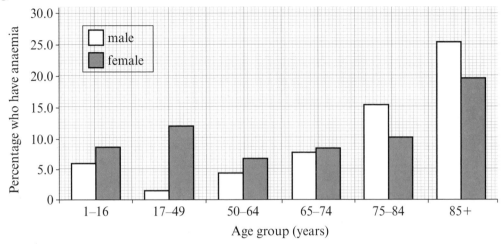

Age group (years)

(a) Anaemia is a deficiency disease. State what is meant by deficiency disease.

.... **(1 mark)**

(b) Describe how the incidence of anaemia changes with age in males and females.

> Be sure to describe the trends in both males and females. Also, you are asked to describe – so do not try to explain!

....

....

....

.... **(4 marks)**

2 The table shows the height and mass of three people.

Subject	Mass (kg)	Height (m)	BMI
person A	80	1.80	24.7
person B	90	1.65	
person C	95	2.00	

(a) Complete the table by calculating the BMI for each person.   **(2 marks)**

(b) Identify the person who is obese.

.... **(1 mark)**

3 Explain how measuring waist : hip ratio is better than BMI when predicting risk of cardiovascular disease.

....

....

....

....

.... **(3 marks)**

Cardiovascular disease

1 (a) State **two** ways in which cardiovascular disease may be treated.

....

.... **(2 marks)**

(b) State **two** pieces of advice a doctor might give to a patient with high blood pressure to help them to make lifestyle changes.

....

....

.... **(2 marks)**

(c) Explain why it is more important to prevent cardiovascular disease than to treat it.

....

....

.... **(2 marks)**

2 The table summarises some of the benefits and drawbacks of the different types of treatment for cardiovascular disease.

Guided

Type of treatment	Benefits	Drawbacks
lifestyle changes	no side effects	may take time to work
medication	easier to do than change lifestyle	can have side effects
surgery	once recovered, there are no side effects	
		risk of infection after surgery

Complete the table with benefits and risks of the different types of treatment. **(3 marks)**

3 Angina is chest pain caused by narrowing of the coronary arteries. This can be treated using a stent. A stent is a wire frame that is inserted into the narrowed part of the artery. Angina can also be treated using heart bypass surgery. This is where the narrowed artery is bypassed using a section of artery or vein.

Guided

> Remember that the coronary arteries are in the heart and supply heart muscle. Think about the consequences if they become blocked.

Evaluate the use of surgery to treat angina.

Surgery can help prevent ....but costs more than inserting a

.... and surgery ....

However, ...

....

....

.... **(4 marks)**

Plant defences

1 A student investigated whether garlic contains a substance that kills pathogens. She crushed some garlic in a little water to make a juice. She mixed this juice in a sterile tube with a bacterial culture and a nutrient-rich jelly. She put a lid on the top of the tube and left it for 2 days.

This question is about a practical piece of work. It is a good idea to look at your practical work before the exam to remind yourself of the skills you use when carrying out an experiment.

She then repeated the investigation but used water rather than garlic juice.

(a) State **two** ways in which plants protect themselves from pests and pathogens using physical barriers.

....

.... **(2 marks)**

(b) State **one** way in which plants use substances to protect themselves against pests and pathogens.

.... **(1 mark)**

(c) State the name of the part of the investigation that used water only.

.... **(1 mark)**

(d) Give **two** reasons why a lid was placed over the top of the tubes.

> Guided

To stop other ....

.... **(2 marks)**

(e) State **two** factors, other than those mentioned in the question, that should be kept constant in this investigation.

....

.... **(2 marks)**

(f) State an advantage to the garlic of containing an antibacterial substance.

.... **(1 mark)**

2 (a) State two different plant defences that may reduce the number of caterpillars on a plant.

....

.... **(2 marks)**

(b) Substances that plants produce for defence can be used as medicines. Aspirin, from the willow tree, is used to treat pain and fever. Give **two** ways that substances produced by plants for defence can be used as medicines.

....

....

....

.... **(2 marks)**

Plant diseases

1 A farmer noticed that plants growing in one field had yellow leaves that curled at the edges and some of them had brown spots. He consulted an agricultural adviser for help.

(a) State **one** environmental factor that can cause plant disease.

.... **(1 mark)**

(b) State **two** visible symptoms that may be a sign of plant disease.

....

.... **(2 marks)**

(c) State why a farmer cannot rely on visible symptoms to know which pathogen has infected their crop.

.... **(1 mark)**

(d) The farmer said that most of the plants in the field seemed to be affected. Explain why the adviser asked the farmer to send:

(i)   samples of soil from several different places in the field

....

.... **(2 marks)**

(ii) a sample of one of the affected plants.

....

.... **(2 marks)**

2 Chronic oak decline is a disease that has affected oak trees throughout the UK since the early twentieth century. It is thought to be caused by fungal infection. More recently, acute oak decline has been identified in south-east England and the Midlands. This is thought to be caused by bacterial infection. Describe a series of tests that scientists could carry out to identify the type of infection.

> **Guided**

Take samples from affected trees and examine with a microscope for signs of

....

....

....

....

....

....

.... **(4 marks)**

Extended response – Health and disease

The widespread availability of antibiotics and vaccines has transformed the treatments of infectious diseases in the last hundred years.

Discuss the relative advantages and disadvantages of antibiotics and vaccines in the treatment of infectious diseases.

> You will be more successful in extended response questions if you plan your answer before you start writing.
>
> You don't need to draw a conclusion, but you do need to write a balanced account that covers both advantages and disadvantages of the two treatments.
>
> Don't be tempted to say too much about how each type of treatment works – focus on how effective they are at treating or preventing infectious diseases.

....

....

....

....

....

....

....

....

....

....

....

....

....

.... **(6 marks)**

Photosynthesis

1 Explain why it is that food chains start with plants or algae.

> Think about what a food chain represents. You will need to use terms such as producer and biomass in your answer.

....

....

.... **(3 marks)**

2 (a) Complete the equation to show the reactants and products of photosynthesis.

⟩ **Guided** ⟩   .... + water → .... + ... **(2 marks)**

(b) Explain why photosynthesis is an endothermic reaction.

....

.... **(2 marks)**

3 A student knew that the products of photosynthesis are converted into starch in leaves. She also knew that iodine solution can be used to test for the presence of starch, producing a blue-black colour. She devised the following experiments to investigate photosynthesis.

Experiment 1

- A plant was kept in the dark for 48 hours to remove all starch from the leaves
- Some of the leaves were covered in foil
- The plant was then placed on a sunny windowsill all day
- At the end of the day two leaves were tested for starch
- One leaf had been covered in foil and did not produce a blue-black colour when tested with iodine
- The other leaf had not been covered in foil and produced a blue-black colour when tested with iodine

---

Experiment 2

- A plant with variegated (partially green, partially white) leaves was kept in the dark for 48 hours to remove all starch from the leaves
- The plant was placed on a sunny windowsill all day
- At the end of the day one leaf was tested for starch
- Only the green parts of the leaf gave a positive test for starch

---

(a) Explain what you can conclude about the requirements for photosynthesis from Experiment 1.

....

.... **(2 marks)**

(b) Explain what you can conclude about the requirements for photosynthesis from Experiment 2.

....

.... **(2 marks)**

Limiting factors

1 (a) Name **one** factor other than carbon dioxide concentration and light intensity that limits the rate of photosynthesis.

.... **(1 mark)**

(b) Describe how you could measure the rate of photosynthesis using algal balls.

....

....

.... **(3 marks)**

2 The graph shows how the rate of photosynthesis changes with light intensity. The data shows the rate at three different concentrations of carbon dioxide.

(a) Describe how increasing the concentration of carbon dioxide changes the rate of photosynthesis.

.... **(1 mark)**

(b) Commercial growers often increase the concentration of carbon dioxide in their greenhouses.

Explain how this will increase the yield of crops grown in the greenhouse.

....

.... **(2 marks)**

Guided (c) Explain how the rate of photosynthesis could be increased further.

You could increase the .... as this

would make photosynthesis happen.. **(2 marks)**

3 A farmer notices that when he changes the temperature of his greenhouse from 15 °C to 25 °C the plants grow more quickly. The plants grow at the same speed at 35 °C, but do not grow at all at 45 °C. The farmer knows that photosynthesis uses enzymes. Explain why the growth of the plants changes in this way.

> If you are asked to explain something make sure you do explain rather than just describe.

....

....

....

....

.... **(4 marks)**

 **Practical skills**

Light intensity

1 Some students wanted to investigate how the rate of photosynthesis in pond weed changed with light intensity. They did this by putting a lamp at different distances from some pond weed in a test tube. They counted the number of bubbles produced by the plant. Here is the data they collected.

Distance from lamp in cm	5	10	15	20	25	30
Number of bubbles per minute	124	88	64	42	28	16

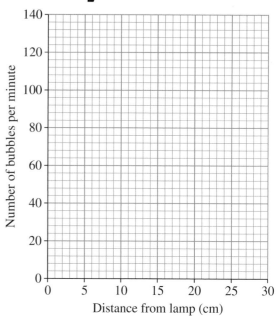

 (a) Plot a graph to show the results in the table. **(2 marks)**

> Mark the points accurately on the grid (to within half a square) using the table of data. Then draw a line of best fit through these points. This line does not have to be straight.

 (b) Use your graph to find the number of bubbles you would expect in 1 minute if the lamp was placed 12 cm from the pond weed.

.... **(1 mark)**

(c) Describe the relationship between light intensity and rate of photosynthesis.

....

.... **(2 marks)**

(d) (i) State **one** safety step you should take.

....

.... **(1 mark)**

(ii) Explain **one** step you should take to ensure your results are reliable.

> In this question, you have to say what the step is and why you need to take that step.

....

....

.... **(2 marks)**

 (e) Describe how you could use a light meter to improve the experiment.

Guided You could use the light meter to measure the .... at each

distance and then plot a graph of ....

.... **(2 marks)**

Specialised plant cells

1 The diagram shows part of a plant tissue specialised for transport.

(a) State the name of this type of tissue.

.... **(1 mark)**

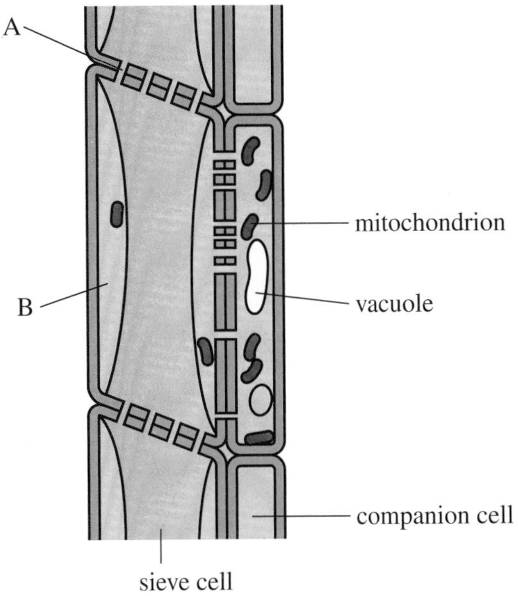

mitochondrion

vacuole

B

A

companion cell

sieve cell

Guided

(b) Explain how the features labelled A and B are adapted to the function of this tissue.

A ....

....

B There is a small amount of cytoplasm so ....

.... **(4 marks)**

(c) Explain why companion cells have many mitochondria.

> Mitochondria supply energy. You need to give this information AND explain why companion cells need lots of energy.

....

....

.... **(2 marks)**

2 (a) State the name of the vessels used to transport water in plants.

.... **(1 mark)**

Guided

(b) Describe **three** ways in which these vessels are adapted for their function.

1 The walls are strengthened with lignin rings to..

....

2 ....

....

3 ....

.... **(3 marks)**

Transpiration

1 A student set up the following experiment to investigate transpiration.

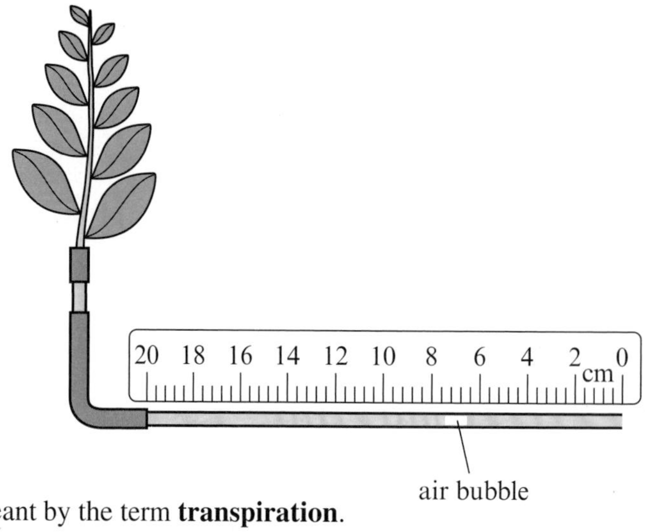

(a) State what is meant by the term **transpiration**.

....

.... **(2 marks)**

(b) State which part of the plant regulates the rate of transpiration.

.... **(1 mark)**

(c) Explain what happens to the air bubble if:

Remember that in an **explain** question you need to say what happens and why.

(i) a fan is started in front of the plant

....

.... **(2 marks)**

(ii) the undersides of the leaves of the plant are covered with grease.

....

.... **(2 marks)**

2 (a) Explain how the guard cells open and close.

....

....

.... **(3 marks)**

(b) The stomata are open during the day but closed at night. Explain why, in very hot weather, plants wilt during the day but recover during the night.

Guided

The stomata are open during the day, so water is lost by .... faster than it can be absorbed by the .... Water is lost from the vacuoles and the plant wilts. At night, the stomata ....

.... **(3 marks)**

Translocation

1 (a) State what is meant by translocation.

....

.... **(1 mark)**

(b) What is the name of the plant tissue responsible for translocation?

☐ **A** phloem

☐ **B** xylem

☐ **C** meristem

☐ **D** mesophyll   **(1 mark)**

2 (a) Describe how radioactive carbon dioxide can be used to show how sucrose is transported from a leaf to a storage organ such as a potato.

Guided

Radioactive carbon dioxide is supplied to the ....

....

....

.... **(3 marks)**

(b) Explain what the effect would be if an inhibitor of active transport were applied to the leaf in this experiment.

> This question requires you to apply knowledge from other parts of the course. Make sure you describe what would happen and give an explanation.

....

....

.... **(2 marks)**

3 The table lists some of the structures and mechanisms involved in movement of water and sucrose in the plant. Put an X in each row of the table to show whether the structure or mechanism is involved in the transport of water or sucrose.

> You might need to revise transpiration on page 72 of the Revision Guide before answering this question.

Structure or mechanism	Transport of water	Transport of sucrose
Xylem		
Phloem		
Pulled by evaporation from the leaf		
Requires energy		
Transported up and down the plant		

(5 marks)

Leaf adaptations

1 The diagram shows a cross-section of part of the leaf of a plant.

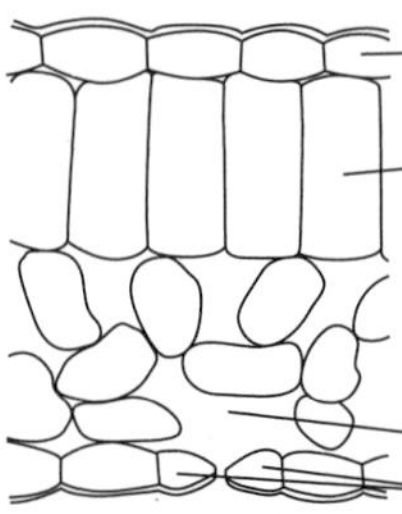

(a) (i) State the name of the structure labelled A.

.... **(1 mark)**

(ii) Explain how structure A allows gas exchange whilst reducing water loss.

epidermal cells

mesophyll cells

B

A

> Think about how the plant balances the need to allow entry of carbon dioxide whilst limiting loss of water.

....

.... **(2 marks)**

(b) Explain how the following are adapted to maximise the rate of photosynthesis.

(i) the epidermis

....

.... **(2 marks)**

(ii) the mesophyll cells

....

.... **(2 marks)**

(c) Explain the adaptation of the part of the leaf labelled B.

....

.... **(2 marks)**

2 Photosynthesis requires light energy, carbon dioxide and water to produce sugars.

> Guided

(a) Explain why plants growing where there is not much light have large leaves.

Large leaves also have a large .... so they can

.... **(2 marks)**

(b) Describe how leaves are adapted to transport water and sugars.

....

....

.... **(2 marks)**

Water uptake in plants

1   Explain the effect on transpiration of the following:

(a) an increase in light intensity

....

.... **(2 marks)**

(b) an increase in temperature

....

.... **(2 marks)**

2   Some students investigated the rate at
    which water evaporated from leaves using
    this apparatus.

    leafy shoot

    The students measured how far the air bubble travelled
    up the capillary tube in 5 minutes with the fan on,
    and with the fan off. They found that the bubble moved
    90 mm with the fan off and 130 mm with the fan on.

    rubber tube

    capillary tube

    air bubble

    water

(a) Explain the results the students collected.

....

....

.... **(3 marks)**

(b) The speed of the fan was increased, and it was found that the rate that the bubble moved did not
     increase. Explain, in terms of the plant's response, why this was the case.

┌───┐
│ In this question, you need to think about what the plant does as the speed of the wind increases. │
└───┘

....

....

....

.... **(3 marks)**

(c) The capillary tube had a diameter of 0.5 mm. Calculate the rate of transpiration in mm^3 / min
     when the fan was off.

> **Guided**

The volume of the tube is calculated using $\pi r^2 l$ ....

....

....

rounded to 1 decimal place = ... **(2 marks)**

Plant adaptations

1   This question is about plants adapted to grow in the rainforest where the ground is very wet and dark.

Guided

(a) Lianas start life in the rainforest canopy and send roots down to the ground. Explain how this adaptation allows them to survive.

The lianas' leaves are high up where there is ....

but the roots are in the ground where it is... **(2 marks)**

(b) The figure shows a leaf of philodendron, a common houseplant that grows wild in rainforests.

Explain **one** adaptation that philodendron has to help it grow in the rainforest.

....

.... **(2 marks)**

2   (a) Pine trees grow in places where the water in the ground is frozen for several months of the year. Unlike deciduous trees, pine trees do not lose their needle-shaped leaves in the winter. Explain how the shape of their leaves allows pine trees to survive in winter without losing their leaves.

> Think about the effect on the tree if the water in the ground is frozen.

....

.... **(2 marks)**

(b) The oleander is a shrub that grows in dry conditions around the Mediterranean. Explain how the following adaptations help oleander to survive.

(i)   a thick, waxy cuticle on the upper surface of the leaves

....

.... **(2 marks)**

(ii) stomata that are sunken in deep pits lined with hairs

....

.... **(2 marks)**

3   Big sagebrush is a desert plant that grows in an area that has some rainfall in spring. Small leaves grow on the plant all year round. Large leaves grow in the spring but fall off in the summer. Explain how these adaptations help big sagebrush to survive.

> Think about why the plant has large leaves in the spring, and the advantages of having small leaves at other times of the year.

....

....

.... **(3 marks)**

Plant hormones

1   The diagram shows how a plant shoot responds to light.

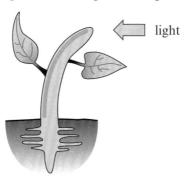

light

(a) Place an X on the diagram where auxins are produced. **(1 mark)**

(b) Place a Y on the diagram to show where auxins travel causing the shoot to bend. **(1 mark)**

(c) Place a Z on the part of the plant that shows positive gravitropism. **(1 mark)**

2   (a) Explain why phototropism is important for shoots.

      Phototropism causes shoots to ....

      This means the leaves are positioned better to..

      .... **(2 marks)**

⟩ **Guided** ⟩

(b) Explain why gravitropism is important for roots.

      ....

      ....

      .... **(2 marks)**

3   Explain how auxins cause the shoot to bend.

> Read the question carefully. You can use the diagram above to help you answer this question if necessary. Think about where auxin is produced and how it causes a plant shoot to bend in a place where it is not produced.

      ....

      ....

      ....

      ....

      ....

      .... **(3 marks)**

Uses of plant hormones

1 (a) Describe **two** ways in which gibberellins can be used to increase crop yields.

....

....

....

.... **(2 marks)**

(b) Explain why fruit growers believe it is worth the expense of spraying flowers with gibberellins.

....

....

....

....

....

.... **(3 marks)**

(c) Many plants, such as hydrangeas, flower in July and August. However, growers at the Chelsea Flower Show in May each year want to display hydrangea plants in full flower. Describe how a grower could produce hydrangea plants flowering in time for the Chelsea Flower Show.

....

....

....

.... **(2 marks)**

2 Describe **two** advantages of using ethene to control fruit ripening after a period in cold storage.

....

....

....

.... **(2 marks)**

3 Describe how selective weedkillers control weeds in fields of crops.

....

....

....

.... **(2 marks)**

4 Explain why rooting powder is used by some plant growers to produce more plants.

....

....

....

.... **(2 marks)**

Extended response – Plant structures and functions

 Plants need to exchange gases with their surroundings. Explain how the need for gas exchange can lead to excess water loss from plants growing in very dry conditions and the adaptations in such plants to help them survive.

> You will be more successful in extended response questions if you plan your answer before you start writing.
>
> This question is about water loss, so make sure you explain how leaves are adapted for gas exchange and how this can lead to water loss.
>
> Then you can explain the adaptations of plants growing in very dry conditions. Make sure you focus on adaptations that help plants survive in dry conditions.

....

....

....

....

....

....

....

....

....

....

....

.... **(6 marks)**

Hormones

Guided

1 (a) Describe how hormones behave like 'chemical messengers'.

Hormones are produced by .... and are released

into the .... They travel round the body until they reach

.... which responds by releasing

.... **(4 marks)**

(b) Describe **two** ways in which hormones and nerves communicate differently.

> Make sure you describe two ways and that they are differences, not similarities.

....

.... **(2 marks)**

2 The diagram shows the location of some endocrine glands in the body. Write in the name of each gland on the corresponding label line.

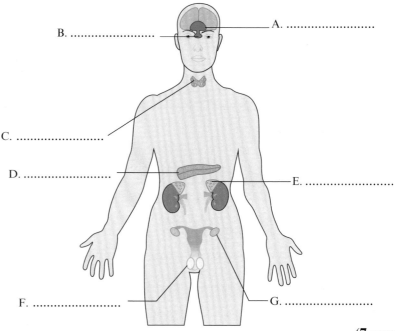

B. ....

A. ....

C. ....

D. ....

E. ....

F. ....

G. ....

(7 marks)

3 Complete the table showing where some hormones are produced and where they have their action.

Hormone	Produced in	Site of action
TRH		pituitary gland
TSH	pituitary gland	
ADH	pituitary gland	
FSH and LH		ovaries
insulin and glucagon		liver, muscle and adipose (fatty) tissue
adrenalin		various organs, e.g. heart, liver, skin
progesterone		uterus
testosterone		male reproductive organs

(4 marks)

Adrenalin and thyroxine

1 (a) A man is walking through a forest at dusk and hears a wolf howl. Explain **two** ways in which adrenalin prepares his body for action in this situation.

....

....

....

....

....

.... **(4 marks)**

(b) Thyroxine controls the resting metabolic rate. Explain how control of thyroxine concentration in the blood is an example of negative feedback.

> Note that in this question you are asked to talk about production of thyroxine, not how it works in controlling metabolic rate. Limit your explanation to the principles of negative feedback. The details will be needed in part (c).

....

....

....

.... **(2 marks)**

(c) Explain how the hypothalamus and pituitary work together to control the amount of thyroxine produced by the thyroid gland.

....

....

....

....

....

.... **(4 marks)**

2 The concentration of thyroxine in the blood is relatively constant but the concentration of adrenalin can change a lot. Explain differences in the pattern of concentration of the two hormones in the blood.

Thyroxine controls the resting metabolic rate so ....

....

Adrenalin is produced in response to ....

....

....

.... **(4 marks)**

The menstrual cycle

1 State **two** of the hormones that control the menstrual cycle.

.... **(2 marks)**

2 The diagram below shows the timing of some features in a menstrual cycle.

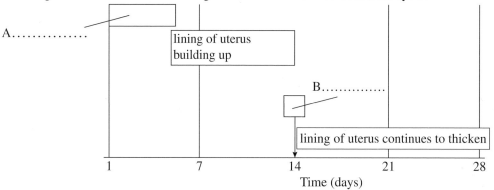

A...............

lining of uterus building up

B...............

lining of uterus continues to thicken

1   7   14   21   28

Time (days)

(a) Fill in the two missing labels, A and B, on the diagram.   **(2 marks)**

(b) Mark with an X on the diagram the point at which fertilisation is most likely to occur.   **(1 mark)**

(c) Describe what happens during days 1–5 of the cycle.

....

....

.... **(2 marks)**

3 (a) Explain how hormonal contraception prevents pregnancy.

> **Guided**

Pills, implants or injections release hormones that prevent ....

and thicken ...., preventing .... from passing.   **(3 marks)**

(b) The table shows the success rates of different methods of contraception.

Method of contraception	Success rate (% of pregnancies prevented)
hormonal pill or implant	> 99%
male condom	98%
diaphragm or cap	92 – 96%

(i) Explain why the actual success rate can sometimes be lower than the figures shown.

....

.... **(2 marks)**

(ii) Evaluate the different methods of contraception for effectiveness in preventing pregnancy as well as protection against STIs.

> Be sure to answer both parts of this question. You might need to review the section on STIs on page 52.

....

....

....

.... **(3 marks)**

Control of the menstrual cycle

1 For each of the following, state where it is released and the target organ.

> This can be confusing, because some hormones are made in the pituitary and act on the ovary while others are made in the ovary and act on the pituitary. Make sure you get these the right way round.

(a) LH

....

.... **(2 marks)**

(b) progesterone

....

.... **(2 marks)**

(c) oestrogen

....

.... **(2 marks)**

2 The diagram shows how the levels of four hormones change during the menstrual cycle.

(a) (i) Explain why the level of FSH rises during the first 7 days of the cycle.

Levels of progesterone are low which

....

....

.... **(2 marks)**

(ii) Explain what causes ovulation to occur around day 14 of the cycle.

....

.... **(2 marks)**

(iii) Explain whether the female is pregnant.

....

.... **(2 marks)**

(b) Hormonal contraception uses a progesterone-like hormone. Explain how this can prevent fertilisation.

....

....

.... **(3 marks)**

(c) At the same time that the level of LH increases there is an increase in the woman's body temperature. Explain how a woman could use this to increase her chance of conceiving a child.

....

.... **(2 marks)**

Assisted Reproductive Therapy

Guided

1 (a) Clomifene therapy can be useful for women who have difficulty conceiving a child. Explain how clomifene increases the chance of pregnancy in women who rarely release an egg during their menstrual cycles.

Clomifene helps increase concentration of...

so stimulates ....

.... **(2 marks)**

(b) Explain why clomifene therapy on its own cannot help a woman with blocked oviducts to become pregnant.

> Think about what clomifene does and why it might not help a woman with blocked oviducts.

....

....

.... **(2 marks)**

2 IVF can be used to help couples who are unable to conceive a child because of problems such as blocked oviducts in the woman or if the man produces few healthy sperm cells.

(a) Explain why the woman is given injections of the hormone FSH at the start of IVF treatment.

....

....

.... **(2 marks)**

(b) Describe the steps that take place after egg cells from the woman and sperm cells from the man are obtained.

....

....

.... **(2 marks)**

(c) Explain why IVF can be useful for couples who risk passing on genetic disorders even if they are able to conceive normally.

....

....

.... **(2 marks)**

3 In 2010, 45 250 women underwent IVF treatment in the UK. Of these women, 12 400 were successful in having a child. The cost of a cycle of IVF treatment is £2500.

Use this information, and your own knowledge, to describe the benefits and drawbacks of IVF treatment.

....

....

....

....

.... **(4 marks)**

Homeostasis

1 State what is meant by homeostasis.

....

....

.... **(2 marks)**

2 (a) State the location in the body of the thermoregulatory centre.

.... **(1 mark)**

   (b) Explain why thermoregulation is important for the enzymes in the body.

....

....

.... **(2 marks)**

   (c) Explain why we shiver if we don't wear enough clothes on a cold day.

> **Guided**

Shivering means energy is released from ....,

which ....

.... **(2 marks)**

3 (a) Explain what is meant by osmoregulation.

....

....

.... **(2 marks)**

   (b) A student had heard that it was important to drink lots of water and so drank two 500 cm^3 bottles of water over a 30-minute period. The student found they had to urinate several times in the following 2 hours. Explain why the student's body had to get rid of water quickly.

....

....

.... **(2 marks)**

4 Suggest an explanation for why osmoregulation does not mean that the body always contains exactly the right amount of water.

> **Suggest** means that you may not have learned this, but you should be able to work it out. Think about how homeostasis works.

....

....

....

....

.... **(3 marks)**

Controlling body temperature

1   The diagram shows some of the structures in the skin involved with regulation of body temperature.

> In both of these questions you need to use the names of the structures (muscle and sweat gland) as well as explaining what each one does.

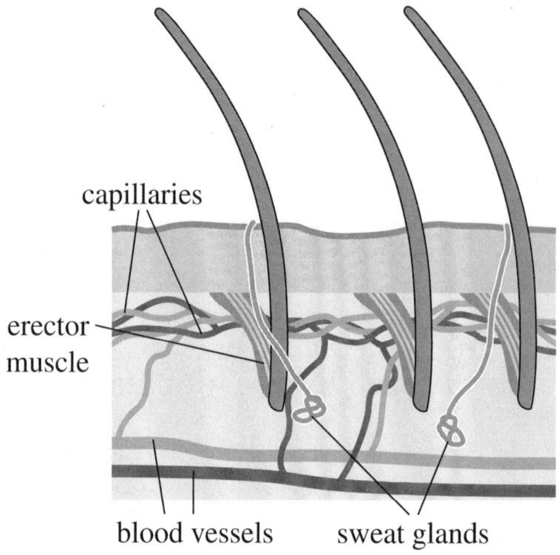

capillaries

erector muscle

blood vessels   sweat glands

(a) Explain how the erector muscles help in control of body temperature.

....

....

....

....

.... **(3 marks)**

(b) Explain how the sweat glands help to cool the body.

....

.... **(2 marks)**

2   An investigation was carried out into the effect of exercise on core (internal) body temperature and the temperature of the skin's surface. The graph shows the results of this investigation.

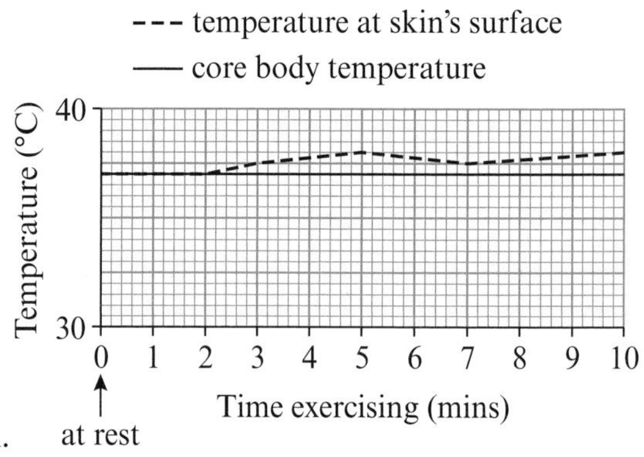

--- temperature at skin's surface
—— core body temperature

Temperature (°C)

40

30

0 1 2 3 4 5 6 7 8 9 10

Time exercising (mins)

↑ at rest

(a) Compare the trends in the data shown by the graph.

....

.... **(2 marks)**

> Guided

(b) Explain how blood vessels in the skin cause a rise in the temperature of the skin's surface during exercise.

> It is important to use scientific words such as 'radiation' and 'dilate'.

When the temperature rises blood vessels ....

There is greater ....

....

This means that more ....

.... **(3 marks)**

(c) Explain why a person with fair skin looks even more pale than usual when they are very cold.

....

.... **(2 marks)**

Blood glucose regulation

1   The table shows the events that happen after a person eats a meal. Complete the table to show the order in which the events take place.

Event	Order
Pancreas increases secretion of insulin	
Blood glucose concentration falls	
Blood glucose concentration rises	1
Insulin causes muscle and liver cells to remove glucose from blood and store it as glycogen	
Pancreas detects rise in blood glucose concentration	

(3 marks)

2   (a) The diagram shows how blood glucose concentration is regulated. Use the information below to fill in the corresponding boxes on the diagram.

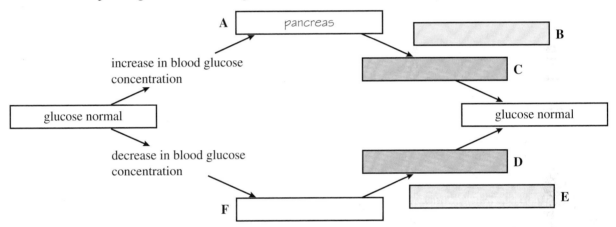

A and F – this gland releases hormones that regulate blood glucose concentration

B – the name of a hormone involved in regulating blood glucose concentration

C and D – the target organ for the hormones involved in blood glucose regulation

E – the name of a hormone involved in regulating blood glucose concentration   **(4 marks)**

(b) Explain what happens in the liver when:

   (i)   the blood glucose concentration rises above normal

   ....

   ....

   .... **(2 marks)**

   (ii) the blood glucose concentration drops below normal.

   ....

   ....

   .... **(2 marks)**

Diabetes

1 (a) The graph shows the percentage of people in one area of the USA in the year 2000 who have Type 2 diabetes, divided into groups according to body mass index.

Describe the link between BMI and the percentage of Type 2 diabetes.

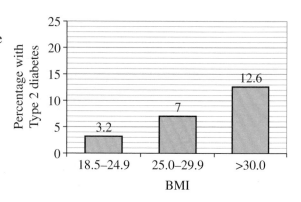

....

.... **(2 marks)**

(b) Two 45-year-old males from the area of the USA studied in part (a) wanted to estimate their chances of developing Type 2 diabetes.

(i) George was 180 cm tall and had a mass of 88 kg. Calculate his BMI and use this to evaluate his risk of developing Type 2 diabetes.

.... **(3 marks)**

(ii) Donald had a waist measurement of 104 cm and a hip measurement of 102 cm. The World Health Organization classes a waist : hip ratio of >0.9 as obese. State and explain whether Donald has an increased risk of developing Type 2 diabetes.

....

....

.... **(3 marks)**

2 (a) Explain how helping people to control their diets might help to reduce the percentage of people in the population who have diabetes.

> Guided

Controlling diets will help to ....

Fewer obese people means .... **(2 marks)**

(b) (i) Explain why people with Type 1 diabetes are treated with insulin but most people with Type 2 diabetes are not.

....

.... **(2 marks)**

(ii) Explain why a person with Type 1 diabetes will sometimes wait to see how large a meal is before deciding how much insulin to inject.

....

.... **(2 marks)**

The urinary system

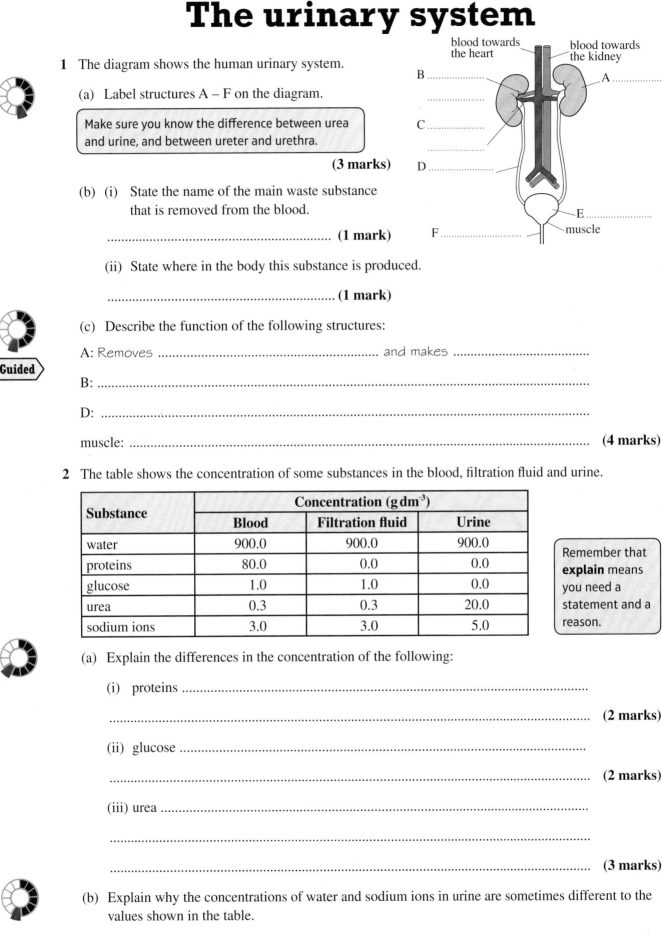

1 The diagram shows the human urinary system.

(a) Label structures A – F on the diagram.

> Make sure you know the difference between urea and urine, and between ureter and urethra.

(3 marks)

blood towards the heart
blood towards the kidney
B ....
....
A ....
C ....
....
D ....
E ....
F ....
muscle

(b) (i) State the name of the main waste substance that is removed from the blood.

.... **(1 mark)**

(ii) State where in the body this substance is produced.

.... **(1 mark)**

Guided

(c) Describe the function of the following structures:

A: *Removes* .... *and makes* ....

B: ....

D: ....

muscle: .... **(4 marks)**

2 The table shows the concentration of some substances in the blood, filtration fluid and urine.

Substance	Concentration (g dm^{-3})		
	Blood	**Filtration fluid**	**Urine**
water	900.0	900.0	900.0
proteins	80.0	0.0	0.0
glucose	1.0	1.0	0.0
urea	0.3	0.3	20.0
sodium ions	3.0	3.0	5.0

> Remember that **explain** means you need a statement and a reason.

(a) Explain the differences in the concentration of the following:

(i) proteins ....

.... **(2 marks)**

(ii) glucose ....

.... **(2 marks)**

(iii) urea ....

....

.... **(3 marks)**

(b) Explain why the concentrations of water and sodium ions in urine are sometimes different to the values shown in the table.

....

....

.... **(2 marks)**

The role of ADH

1 (a) (i)   State where in the body ADH is produced.

.... **(1 mark)**

(ii) State the precise target for ADH.

> The question says 'precise' so 'kidney' will not be enough for the mark.

.... **(1 mark)**

(iii) Explain the effect ADH has on its target.

....

....

....

.... **(2 marks)**

2 A student took part in a five-mile run on a hot day. He did not take any fluids with him to drink.

> Guided

(a) Later in the day the student needed to urinate. Describe how his urine would differ from how it was the previous day when it was cool and he didn't go for a run.

The volume of urine would be ....

and the urine would be more .... **(2 marks)**

(b) Explain how ADH was involved in the control of the amount of water in the student's blood during the run.

> This question is worth four marks, which means that you are going to have to provide quite a lot of detail in your answer.

....

....

....

....

....

....

....

.... **(4 marks)**

Kidney treatments

1 (a) Explain why a person with kidney failure will have to receive kidney dialysis every few days.

> Although you only need to be able to describe treatments for kidney failure, you may have to use earlier knowledge to answer a question like this.

....

.... **(2 marks)**

(b) Describe the function of the partially permeable membrane in a kidney dialysis machine.

....

.... **(2 marks)**

Guided

(c) The table compares the concentrations of some substances during kidney dialysis.

	A	B	Comparison
Concentration of glucose in:	dialysis fluid at start of dialysis	blood at start of dialysis	same
Concentration of glucose in:	dialysis fluid at end of dialysis	blood at end of dialysis	
Concentration of urea in:	dialysis fluid at start of dialysis	blood at start of dialysis	
Concentration of urea in:	blood at start of dialysis	blood at end of dialysis	

Complete the table by writing in whether concentration A or B is higher or if they are the same.

(4 marks)

(d) Explain why it is necessary to keep the dialysis fluid flowing through the machine during kidney dialysis.

....

.... **(2 marks)**

2 Kidney failure often happens as a result of diabetes. What would happen to the concentration of glucose in the blood of a person with diabetes during kidney dialysis? Give a reason for your answer.

....

.... **(2 marks)**

3 Organ donation is a better treatment for most people with kidney failure.

(a) Describe how a kidney transplant is carried out.

....

.... **(2 marks)**

(b) Explain why some people have to wait much longer than others for a suitable donor.

....

.... **(2 marks)**

Extended response – Control and coordination

Compare how Type 1 and Type 2 diabetes are caused and how they are treated.

> You will be more successful in six-mark questions if you plan your answer before you start writing.
>
> Make sure that you cover the causes of each type of diabetes and link this to type of treatment.

....

....

....

....

....

....

....

....

....

....

....

....

....

.... **(6 marks)**

Exchanging materials

1 Substances are transported into and out of the body. Describe where and why the following substances are removed from the blood stream.

(a) Water ....

.... **(2 marks)**

(b) Urea ....

.... **(2 marks)**

2 Humans and other mammals need to exchange gases with their environment. Describe where and why this exchange happens.

....

....

.... **(3 marks)**

3 Absorption of digested food molecules takes place in the small intestine. The small intestine has a surface adapted to assist this process.

> Guided

(a) Describe how the small intestine is adapted to help to absorb food molecules.

The surface of the small intestine is covered with .... These help

by increasing .... **(2 marks)**

(b) Explain why the structures described in part (a) have thin walls.

....

.... **(2 marks)**

4 The diagram shows a flatworm and an earthworm.

> Guided

The two worms are similar in size. Explain why the flatworm does not have an exchange system or a transport system whereas the earthworm has a transport system (heart and blood vessels).

The flatworm is very flat and thin which means it has a large ....

....

....

....

....

.... **(4 marks)**

Alveoli

1 (a) Describe how gas exchange takes place in the lungs.

Oxygen diffuses from .... into ....

Carbon dioxide diffuses from .... into .... **(2 marks)**

(b) State and explain **two** ways in which the structure of the alveoli is adapted for efficient gas exchange.

Millions of alveoli create a large ....

for the .... of gases. Each alveolus is closely associated with

a .... Their walls are one ....

This minimises the .... distance. **(4 marks)**

2 Explain the importance of continual breathing and blood flow for gas exchange.

....

....

....

.... **(2 marks)**

3 Emphysema is a type of lung disease where elastic tissue in the alveoli breaks down. The figure shows the appearance of an alveolus damaged by lung disease compared with a healthy alveolus.

Healthy alveolus   Alveolus damaged by lung disease

Explain how emphysema affects the person.

Think about what effects the changes in emphysema would have on gas exchange and how this would then affect the person.

....

....

....

....

....

.... **(3 marks)**

Rate of diffusion

1 Lungs are an exchange surface that allow gases to be exchanged between the blood and air.

  (a) Describe **three** factors that increase the effectiveness of the lungs as an exchange surface.

    ....

    ....

    .... **(3 marks)**

  (b) For each factor you have mentioned in (a), describe one way in which the lungs are adapted to increase their effectiveness as an exchange surface.

    > Make sure you link each example to each factor you have mentioned.

    ....

    ....

    .... **(3 marks)**

2 Fick's law can be used to calculate the rate of diffusion across an exchange surface.

  (a) Pulmonary fibrosis is caused by damage to the alveoli. It leads to thick scar tissue forming in the lungs. The scar tissue in a person with pulmonary fibrosis has increased the thickness of the membranes in the lungs by a factor of 3. Use Fick's law to describe what would happen to the rate of gas diffusion in this person.

> Fick's law is: rate of diffusion $\propto \dfrac{\text{surface area} \times \text{concentration difference}}{\text{thickness of membrane}}$

    ....

    ....

    .... **(2 marks)**

  (b) Explain what symptoms you would expect in a person with pulmonary fibrosis.

    ....

    ....

    .... **(2 marks)**

Guided

  (c) Coronary heart disease reduces the flow of blood through the lungs. Explain why patients with coronary heart disease have similar symptoms to those you have described in your answer to part (b).

    Reduced blood flow would reduce the...

    in the .... and so less ....

    .... **(2 marks)**

Blood

1 (a) Explain why red blood cells contain large amounts of haemoglobin.

....

....

.... **(2 marks)**

(b) Explain **two** other ways in which the structure of red blood cells is related to their function.

> Make sure that you give two features of red blood cells and, for each one, relate the structure to the function.

....

....

....

....

....

.... **(4 marks)**

2 Describe **one** way in which blood plasma transports substances.

Dissolved substances such as ....

are transported ....

.... **(2 marks)**

3 Explain how platelets help to protect the body from infection.

....

....

....

....

.... **(3 marks)**

4 White blood cells usually make up about 1% of the blood and include lymphocytes and phagocytes.

(a) Explain why the number of lymphocytes increases during infection.

....

....

....

.... **(3 marks)**

(b) Describe how phagocytes help protect the body.

....

....

.... **(2 marks)**

Blood vessels

1 (a) Describe the structure of an artery.

> **Guided**

An artery has .... walls. These walls are composed of two types of fibres:

.... tissue and .... fibres.   **(3 marks)**

(b) Explain how the structure of the artery wall makes blood flow more smoothly in arteries.

....

....

....   **(2 marks)**

2 Blood needs to penetrate every organ in the body. This is made possible by capillaries.

(a) Describe how the capillaries are adapted for this function.

> Make sure you describe here and save the explanation for part (b).

....

....

....   **(2 marks)**

(b) Explain how the features you have described are important for the function of capillaries.

....

....

....   **(2 marks)**

3 (a) Veins carry blood away from the organs of the body to the heart.

(i) Explain why there is a difference in the thickness of the walls of arteries and veins.

....

....

....   **(2 marks)**

(ii) Explain how muscles and valves work together to help to return blood to the heart.

....

....

....   **(2 marks)**

(b) A nurse taking blood from a patient will insert a needle into a vein.

Explain why blood is taken from veins, not from arteries.

> There are two possible answers here – you need to consider either the structure of the different blood vessels, or else the way in which each transports the blood they contain.

....

....

....   **(2 marks)**

The heart

1 The heart is connected to four major blood vessels. Describe where each vessel carries blood. The first one has been done for you.

aorta   carries blood from heart to body

pulmonary artery   carries blood from .... to ....

pulmonary vein   carries blood from .... to ....

vena cava   carries blood from .... to .... **(4 marks)**

2 (a) Explain why the heart consists mostly of muscle.

....

.... **(2 marks)**

(b) Describe the route taken by blood through the heart from the vena cava to the aorta.

....

....

.... **(3 marks)**

3 The diagram shows a section through the human heart.

Remember that the heart is drawn and labelled as if you are looking at the heart in someone's body. So the right side of the heart is actually on the left side of the page!

(a) State the name of the part of the heart labelled A and describe its function.

....

.... **(2 marks)**

(b) Explain the function of the part labelled B.

....

.... **(2 marks)**

(c) Explain why the muscle at C needs to be thicker than on the other side of the heart.

....

....

.... **(3 marks)**

Aerobic respiration

1 Read the following passage and answer the questions that follow.

> Aerobic respiration happens in muscle cells in the body. The muscle cells are surrounded by blood vessels. The substances needed for respiration are transferred to the muscle cells by diffusion, and the waste products are removed.

(a) Name the substances needed for respiration in muscle cells.

.... and .... **(2 marks)**

(b) State the meaning of the term **diffusion**.

Guided

Diffusion is the movement of substances from .... to .... concentration.

(1 mark)

2 (a) State the location in the cell where most of the reactions of aerobic respiration occur.

.... **(1 mark)**

(b) Explain how cellular respiration helps maintain the body temperature.

....

....

.... **(2 marks)**

(c) State **one** way that animals use energy from respiration, other than to maintain their body temperature.

....

.... **(1 mark)**

3 The blood supplies cells with the substances needed for aerobic respiration, as well as removing waste products.

(a) Write a word equation for aerobic respiration.

.... **(1 mark)**

(b) State the name of the smallest blood vessels that carry blood to the respiring cells.

.... **(1 mark)**

4 (a) Explain why all organisms respire continuously.

....

.... **(2 marks)**

(b) Plants can use energy from sunlight in photosynthesis. Explain why plants also need to respire continuously.

> Photosynthesis uses light energy in production of glucose; it does not release energy that can be used in other processes. Think about why plants need energy from respiration.

....

.... **(2 marks)**

Anaerobic respiration

1 Humans can respire in two ways: using oxygen (aerobic) and without using oxygen (anaerobic).

> Make sure that you understand what is produced in both aerobic and anaerobic respiration.

(a) Compare the amounts of energy transferred by aerobic and anaerobic respiration.

....

.... **(2 marks)**

(b) Describe the circumstances under which anaerobic respiration occurs.

....

.... **(2 marks)**

2 In track cycling, a 'sprint' event begins with several slow laps in which the riders try to get a tactical advantage. These slow laps are followed by a very fast sprint to the finishing line.

(a) Describe and explain how the cyclists' heart rates change during the course of the race.

....

....

.... **(3 marks)**

(b) After the race the cyclists will cycle on a stationary bicycle for 5–10 minutes. Explain why they do this.

....

.... **(2 marks)**

3 The graph shows how oxygen consumption changes before, during and after exercise. The intensity of the exercise kept increasing during the period marked 'Exercise'.

(a) Explain the shape of the graph during the period marked exercise.

....

....

.... **(3 marks)**

> Guided

(b) Explain the shape of the graph during the period marked recovery.

During exercise there is an increase in the concentration of ....

....

....

.... **(2 marks)**

 **Practical skills** # Rate of respiration

1   The diagram shows a respirometer used to investigate the rate of respiration in germinating peas.

> This is one of the core practicals so you should be able to answer questions on the apparatus.

syringe containing air

3-way tap

ruler

capillary tubing   blob of liquid

respiring peas
wire gauze
water bath
potassium hydroxide

State the role of the following and give a reason for your answer.

(a)   the water bath

....

....

.... **(2 marks)**

> **Guided**

(b)   the potassium hydroxide

Absorbs carbon dioxide produced by the .... so that

....

.... **(2 marks)**

(c)   the tap and syringe containing air

....

....

.... **(2 marks)**

2   (a)   Explain how the apparatus allows you to measure the rate of respiration in the seeds.

> The movement of the liquid blob gives you information about the uptake of oxygen. You need to state this, explain how you measure the movement and then how you calculate the rate of respiration.

....

....

.... **(3 marks)**

(b)   Describe how you would use the apparatus to investigate the effect of temperature on the rate of respiration in peas.

....

....

.... **(3 marks)**

Changes in heart rate

1 Cardiac output can be calculated using the equation:
cardiac output = stroke volume × heart rate.

(a) What is meant by the term **stroke volume**?

.... **(1 mark)**

(b) A man has a heart rate of 60 beats/minute and an average stroke volume of 75 cm³.

(i) Calculate his cardiac output. Show your working out and give the correct units.

Cardiac output .... **(3 marks)**

(ii) Explain the change in cardiac output when the man starts to exercise.

....

....

.... **(3 marks)**

2 The graph shows the pulse rate of an athlete at rest, and after 5 minutes of different types of exercise.

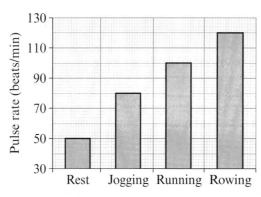

Remember to show all your steps in the calculation.

(a) Calculate the percentage increase in pulse rate between jogging and running.

100 beats/min – 80 beats/min = .../min

(............/80) x 100 =

Percentage increase .... **(2 marks)**

(b) State why the pulse rate is highest when the athlete is rowing.

....

.... **(1 mark)**

(c) The pulse is a measure of heart rate. At rest, the cardiac output of the athlete is 4000 cm³/min. Calculate the stroke volume, in cm³, of the athlete at rest.

Stroke volume .... cm³   **(2 marks)**

Extended response – Exchange

The diagram shows the main features of the human heart and circulatory system.

Describe the journey taken by blood around the body and through the heart, starting from when it enters the right side of the heart. In your answer, include names of major blood vessels and chambers in the heart.

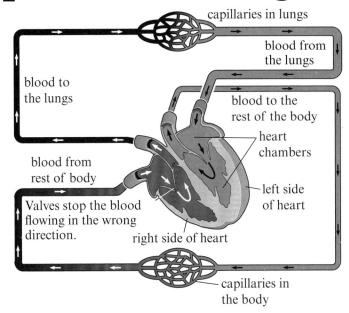

You will be more successful with extended writing questions if you plan your answer before you start writing.

You do not need to explain how the different components of the heart and circulatory system work. It may help your plan if you follow the blood around the diagram with a finger, writing the name of each blood vessel or chamber in order as you go. You do not need to identify any blood vessels in the 'rest of the body' other than the aorta.

....

....

....

....

....

....

....

....

....

....

.... **(6 marks)**

Ecosystems and abiotic factors

1 Draw lines to connect each term with its definition.

Term	Definition
Community	A single living individual
Organism	All the living organisms and the non-living components in an area
Population	All the populations in an area
Ecosystem	All the organisms of the same species in an area

(4 marks)

2 A student surveyed the distribution of a species of lichen growing on the trunk of a tree. She used a small quadrat to measure the percentage cover by these lichens on the south and north facing sides of the tree.

	Light intensity / lux			
	Reading 1	**Reading 2**	**Reading 3**	**Mean**
South side	275.5	368.1	326.8	
North side	195.7	282.1	205.1	

Percentage cover / %											
South side						**North side**					
1	2	3	4	5	Mean	1	2	3	4	5	Mean
48	20	28	92	8	39	4	4	4	4	6	4

Guided

(a) Complete the upper table to show the mean light intensity for each side. **(2 marks)**

(275.5 + 368.1 + 326.8)/3 = ...

(b) The student concluded that the lichen was better adapted to conditions on the south side. Justify her conclusions.

> In this part of the question you are only asked to say whether she was right and, if so, why. Don't try to explain her results.

....

....

....

.... **(2 marks)**

(c) Light intensity is an abiotic factor. Explain one other abiotic factor that might be responsible for the different distribution of lichen.

....

....

....

.... **(2 marks)**

Biotic factors

1 Meerkats are animals that live in packs and are found in the desert areas of southern Africa. The pack of meerkats is led by a dominant pair of meerkats, known as the alpha male and female.

> In this question, you will be asked to think about aspects of the behaviour of the meerkats. Remember to link your answer to the ideas that these animals will compete with each other for resources.

(a) State what the term **biotic factors** means.

.... **(1 mark)**

(b) Only the alpha male and alpha female breed. Suggest an explanation for why younger male meerkats will often try to fight the alpha male.

....

.... **(2 marks)**

(c) When meerkat packs become very large, they often split into smaller packs. The new pack will often move some distance from the original pack. Explain the reasons why a large pack may need to split up.

> Make sure that you know what the command words mean. **Explain** means give a reason why. **Suggest** means you need to apply your knowledge to a new situation. **Describe** means say what is happening.

....

....

.... **(2 marks)**

2 The drawing shows a male peacock.

State and explain **one** adaptation, seen in the diagram, that helps the peacock attract a mate.

....

....

....

.... **(3 marks)**

3 The diagram shows a cross-section through a tropical rainforest.

> Guided

(a) Some trees are called emergent. They break through the rest of the rainforest canopy. Explain the advantage to these trees of emerging from the canopy.

The trees emerge through the canopy

to get .... for more .... **(2 marks)**

emergent layer

canopy layer

(b) The soil in a rainforest is often poor as the minerals are washed away (leached). Suggest an explanation of how trees in the rainforest may adapt to respond to a leached soil.

....

.... **(2 marks)**

Parasitism and mutualism

1 Compare and contrast parasitism and mutualism.

....

....

....

.... **(3 marks)**

> You could get a question like this as an extended response question in an exam, where you would probably be asked to give examples to illustrate your answer. Here it is enough to say how the relationships are similar and how they are different.

2 Fleas are small insects that feed on the blood of animals.

(a) Describe what each organism gets out of this relationship.

Fleas: ....

....

Animals: ....

.... **(2 marks)**

(b) Explain what type of feeding relationship exists between fleas and animals.

....

....

.... **(2 marks)**

3 Cleaner fish are small fish that feed on parasites on the skin of sharks. Describe how the cleaner fish and the sharks benefit from a mutualistic relationship.

⟩ **Guided** ⟩

Cleaner fish get food by..

....

This helps the shark because .... **(2 marks)**

> An exam question may ask you about the benefits to one organism or to both. Make sure you read the question carefully!

4 The scabies mite is a tiny insect that burrows into human skin and lays its eggs. Infection by the scabies mite causes severe itching and a lumpy, red rash that can appear anywhere on the body. Explain why the scabies mite is a parasite and not a mutualist.

> This type of question is expecting you to apply your understanding of science to a situation that you may not be familiar with. You will have been taught about organisms that behave in a similar way to the scabies mite – use what you know about these organisms but apply it to the scabies mite.

....

.... **(2 marks)**

Fieldwork techniques

1 A gardener goes into his garden every night at 7pm and counts the number of slugs in the same 1 m² area of his flower bed. He records his results in a table.

Day	Monday	Tuesday	Wednesday	Thursday	Friday	Saturday	Sunday
Number of slugs	11	12	7	12	8	8	12

(a) Describe how the gardener could make sure the 1 m² area of the flower bed was chosen at random on the first day.

....

.... **(2 marks)**

> **Guided**

(b) Why does the gardener use the same area each time?

Using the same area means that his experiment is .... **(1 mark)**

(c) Describe **one** way in which the gardener could improve the repeatability of the data that he collected.

....

.... **(2 marks)**

2 A class is investigating the number of clover plants on a football pitch. The pitch measures 100 m by 65 m. The class wants to find the total number of clover plants in the field. The teacher gives the class a 1 m × 1 m quadrat.

$$\text{mean number of plants} = \frac{\text{total number of plants in all quadrats}}{\text{number of quadrats}}$$

(a) Explain how the class can use the quadrat to estimate the mean number of clover plants in a 1 m² area.

....

.... **(2 marks)**

(b) The class finds that the mean number of clover plants in an area of 1 m × 1 m is 7. Estimate the number of clover plants on the whole football pitch.

....

....

.... **(3 marks)**

3 Describe how you would use a belt transect to investigate the distribution of broad-leaved plants growing alongside a path that started at a road, crossed a small field and entered a wood.

Make sure you describe use of quadrats, the measurements you would take and what you would record.

....

....

....

.... **(3 marks)**

Organisms and their environment

1 Limpets are animals that have a shell and live on rocks that are underwater some or all of the time. They can be found in the sea, or in rock pools on the beach. A scientist is investigating the distribution of limpets on the beach.

(a) Explain how the scientist could use a transect to investigate the distribution of limpets.

....

....

.... **(3 marks)**

The scientist sets up three different transects and measures the numbers of limpets on each one. His data is shown in the table.

Distance from sea in metres	Number of limpets			Mean number of limpets
	Transect 1	Transect 2	Transect 3	
0.5	20	23	20	21
1.0	18	16	17	17
1.5	13	13	13	13
2.0	10	8	9	
2.5	5	6	4	5

(b) Calculate the mean number of limpets at 2.0 m from the sea in this investigation.

.... **(2 marks)**

(c) What conclusion can be made from his investigation?

> Your conclusion should describe how the distribution of limpets changes along the transect.

....

.... **(3 marks)**

2 A scientist investigated the distribution of bluebells in a large wood. She started on the edge of the wood, and measured a line going deeper into the wood. Every 2 m into the woodland, she placed a quadrat and counted the number of bluebells in the quadrat. She also measured the light intensity at each quadrat.

(a) Describe **one** way the scientist could alter her method to collect more accurate data.

Instead of placing a quadrat every 2 m, the scientist could ....

and use a .... quadrat than before.   **(2 marks)**

(b) The scientist obtained the following data:

Distance from edge of wood in metres	0	2	4	6	8	10	12	14	16
Number of bluebells	0	7	15	22	25	21	16	10	8

Suggest an explanation for these results.

....

....

....

.... **(2 marks)**

Energy transfer between trophic levels

1 The diagram shows a food web.

(a) Identify the producer in this food web.

....

(1 mark)

(b) State the number of trophic levels in this food web.

....

(1 mark)

(c) Explain why it is unusual to have more than this number of trophic levels.

Because there is not enough biomass in ....

....

.... **(2 marks)**

2 The table gives details about the trophic levels in a food chain.

(a) Calculate the biomass at each trophic level. Write your answers in the spaces in the table.

Organism	Energy at each trophic level (J)	Number of organisms	Mass of each organism (kg)	Biomass at each trophic level (kg)
Producers	7550	10 000	0.25	2500
Herbivores	640	200	2.5	
Carnivores	53	10	20	

(1 mark)

(b) Use your calculations to draw a pyramid of biomass for the food chain shown in the table.

> If you are asked to draw a pyramid of biomass, make sure you draw it to scale.

(2 marks)

(c) Calculate the percentage of energy that is transferred from the herbivores to the carnivores.

percentage of energy .... **(2 marks)**

(d) Explain why the amount of energy decreases as it is transferred from one trophic level to the next.

....

....

.... **(2 marks)**

Human effects on ecosystems

1 Fish can be farmed or caught from the wild. State **one** advantage of fish farming, and **one** disadvantage.

Advantage Reduces fishing of ....

Disadvantage ....

.... **(2 marks)**

2 A non-indigenous species is not naturally found in a particular place. For example, the cane toad is a non-indigenous species in Australia that was introduced to control insect pests. State **one** other advantage of introducing a non-indigenous species, and **one** disadvantage.

Advantage ....

....

Disadvantage ....

....

.... **(2 marks)**

3 The graph shows the mass of fertiliser used in the world from 1950 to 2003.

(a) Calculate the percentage increase in fertiliser use from 1950 to 2003.

> Make sure that you read the graph carefully to get the correct figures for your calculation. You will get one mark for showing the correct calculation and one mark for the correct answer.

percentage increase... **(2 marks)**

(b) Suggest an explanation for the change in the mass of fertiliser used worldwide since 1950.

....

....

.... **(2 marks)**

(c) Describe an environmental problem caused by over-use of fertilisers.

....

.... **(2 marks)**

Biodiversity

1 (a) State what is meant by reforestation.

....

.... **(1 mark)**

(b) Describe **two** advantages of reforestation.

....

....

.... **(2 marks)**

2 Explain the importance to humans of conservation.

....

....

.... **(2 marks)**

Guided

3 Yellowstone National Park in the USA is the natural home of many species. Deer eat young trees, stopping them from growing. The population of deer increased so much that Park authorities decided to reintroduce some wolves.

The wolves killed some of the deer for food. The deer moved away from river areas because they were more easily hunted by the wolves there. The wolves also killed coyotes, which are predators that eat rabbits.

The reintroduction of wolves led to major improvements in the biodiversity of the Park. This included increases in the populations of rabbits, bears, hawks and other birds. It also reduced the amount of soil washed into the rivers.

Describe the ways in which the reintroduction of the wolves may have caused the biodiversity to improve.

> This is an example of having to apply knowledge you have learned in this and other units to an unfamiliar situation.

The numbers of trees will increase because ....

This means there will be more food for ....

There will be more rabbits because ....

If there are more rabbits, there will be more food for . ...

More trees also mean ....

.... **(5 marks)**

Food security

1 (a) State what is meant by food security.

.... **(1 mark)**

(b) Explain why increasing human populations increase the need for food security.

> This is not just about the numbers, it is also about living standards.

....

....

.... **(2 marks)**

(c) State what is meant by sustainability.

....

....

.... **(2 marks)**

2 Growing soya to produce biodiesel (a biofuel) can help to reduce use of fossil fuels.

(a) Explain what effect this might have on the food supply for human populations.

....

.... **(2 marks)**

(b) Describe **two** ways in which biodiesel production may not be sustainable.

....

....

.... **(2 marks)**

3 Some people think that intensive farming methods will help to meet growing demand for meat. However, it takes about 7 kg of feed (produced from grain) to produce 1 kg of beef.

> Guided

Explain why some people think the use of intensive farming may not improve food security in a sustainable way.

7 kg of grain will feed more people than 1 kg of beef ....

....

....

....

.... **(3 marks)**

The carbon cycle

Guided

1 Complete the diagram of the carbon cycle by writing the names of the processes in the boxes.

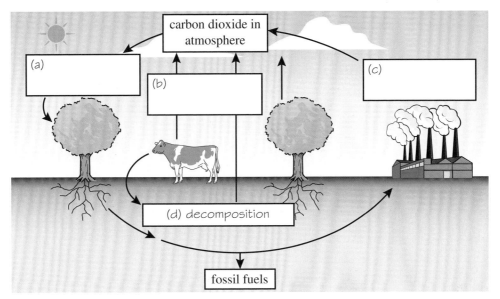

carbon dioxide in atmosphere

(a)

(b)

(c)

(d) decomposition

fossil fuels

(4 marks)

2 Explain why microorganisms are important in recycling carbon in the environment.

....

....

.... **(2 marks)**

> In questions about the carbon cycle, you will be expected to make links between photosynthesis, respiration and combustion, and the amount of carbon dioxide in the air.

3 (a) The diagram shows a fish tank. Explain how carbon is recycled between organisms in the fish tank.

....

....

....

....

....

....

.... **(4 marks)**

(b) Explain why it is important that numbers of both plant and animal populations in the fish tank are kept balanced.

....

....

.... **(3 marks)**

The water cycle

1 (a) Give **three** natural sources of water vapour in the atmosphere.

> The question says 'water vapour' so make sure your answer talks about the formation of water vapour and not about other aspects of the water cycle.

....

....

....

....

....   **(3 marks)**

(b) Describe what happens when water vapour in the atmosphere condenses.

....

....

....

....

....   **(3 marks)**

Guided

2 In parts of California there is a lack of rainfall. Water has been taken from rivers and used to water lawns and golf courses. Some of these areas are suffering from drought and there are now restrictions on the number of days a week golf courses can be watered. Explain why these restrictions have been introduced.

A lot of water evaporates from a golf course so this will lead to...............................

....

....

....   **(3 marks)**

3 Sea water contains too much salt to make it potable (safe to drink). Potable water can be produced from sea water by desalination:

• sea water is evaporated by heating

• water vapour is cooled and condensed.

Give **one** advantage of desalination to people in areas where there is a drought, and **one** disadvantage.

....

....

....

....

....   **(2 marks)**

The nitrogen cycle

The diagram shows how the element nitrogen moves between living organisms and the environment.

1 Bacteria are involved in different stages of the nitrogen cycle. Which is the correct combination of processes involving bacteria?

☐ **A** processes A and B only

☐ **B** processes A, B and C only

☐ **C** processes A, B, C and D only

☐ **D** processes A, B, C, D and E   **(1 mark)**

A Nitrogen fixation in root nodules
B Nitrogen fixation in soil
C Denitrification
D Decomposition
E Absorption

2 (a) Describe the process happening in B.

> Guided

Nitrogen fixation by .... **(1 mark)**

(b) Explain the importance of process E.

....

....

....

....

.... **(3 marks)**

(c) Explain the importance of bacteria in stage C.

....

....

....

.... **(2 marks)**

3 Some farmers use crop rotation, with different crops each year, including a 'green manure' crop such as clover.

Explain the importance of process A in crop rotation.

> You need to say something about how clover is involved in process A as well as its importance for crop plants.

....

....

....

....

.... **(3 marks)**

Pollution indicators

1 The graph shows the oxygen concentration in a river. At one point sewage enters the river. Sewage contains high levels of nitrates.

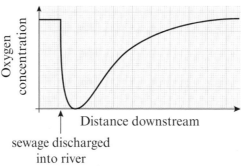

sewage discharged
into river

> **Guided**

(a) Describe the effect of sewage discharge on the oxygen content of the river.

When sewage enters the river, the amount of oxygen ....and

then ....

.... **(2 marks)**

(b) Give **two** examples of organisms that are likely to be found in the river before the sewage enters it.

....

....

.... **(2 marks)**

(c) Sewage contains nitrates. Describe the effect that the nitrates will have on living organisms in the river.

> You need to think through the sequence of events that occurs after eutrophication very carefully. Describe these events in the order in which they happen.

....

....

....

....

....

.... **(4 marks)**

2 Evaluate the use of indicator species to help monitor air quality.

....

....

....

....

....

.... **(4 marks)**

Decay

1 A gardener uses compost in the soil of his garden. The gardener makes the compost himself. He then grows tomatoes in the soil.

Guided

(a) The compost is made from garden waste. What conditions are needed for compost to form from the garden waste?

The conditions needed are oxygen, ...

and .... **(2 marks)**

(b) Explain why the gardener puts the heap in a sunny part of the garden.

....

....

.... **(2 marks)**

(c) Explain why the gardener adds a small amount of water if the heap gets dry.

....

....

.... **(2 marks)**

2 (a) It used to be common for food to be preserved by drying and/or salting. Explain how these help preserve food.

> Think about what causes food to decay and how drying and salting might prevent that happening.

Drying ....

....

Salting ....

....

.... **(3 marks)**

(b) More recent methods of preserving food include refrigeration and packing in nitrogen. Explain how these help preserve food.

Refrigeration ....

....

....

....

Packing in nitrogen ....

....

....

.... **(4 marks)**

Extended response – Ecosystems and material cycles

Explain how fish farming and other human activity has an impact on biodiversity.

> You will be more successful in extended writing questions if you plan your answer before you start writing.
>
> Try to include a number of different examples of how human activity has an impact on biodiversity. Remember that not all human activity is bad for biodiversity; try to think of some examples where human activity can increase biodiversity.

....

....

....

....

....

....

....

....

....

....

....

....

.... **(6 marks)**

Timed Test 1

Time allowed: 1 hour 45 minutes

Total marks: 100

Edexcel publishes official Sample Assessment Material on its website. This practice exam paper has been written to help you practise what you have learned and may not be representative of a real exam paper.

1 Catalase is an enzyme found in many different tissues in plants and animals. It speeds up the breakdown of hydrogen peroxide:

hydrogen peroxide → water + oxygen

A group of students wanted to design an experiment to investigate the amount of catalase in different plant and animal tissues. They knew that when the reaction takes place in a test tube, the oxygen gas given off produces foam. They decided that they could measure the height of the foam in the test tube and use this to estimate the amount of catalase in the different types of tissue.

The group was provided with hydrogen peroxide solution, test tubes and five different plant tissues.

(a) Devise a plan, using the supplied solution and apparatus, to compare the amount of enzyme in different tissues.

(3 marks)

(b) State **two** variables the students should control in the investigation. **(2 marks)**

(c) State one improvement that the students could make that would increase the accuracy of their measurement of enzyme activity. **(1 mark)**

(Total for question 1 = 6 marks)

2 A man has an infection of disease-causing bacteria. He has not been immunised against these bacteria. The graph shows how the number of these bacteria change after a doctor gives the man a 7-day course of antibiotics.

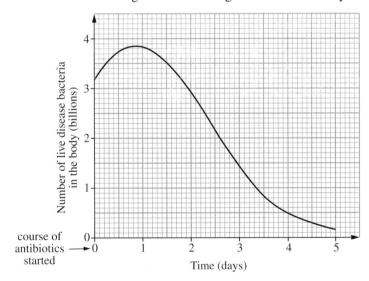

(a) Explain whether the man has a communicable or non-communicable disease. **(2 marks)**

(b) The man started feeling better on day 3. Calculate the percentage decrease in the number of live disease-causing bacteria between the start of the course of antibiotics and the time that the man started to feel better. Give your answer to one decimal place. **(2 marks)**

(c) The doctor suspected that the disease was caused by a bacterium not a virus. Use the information in the graph to explain why the doctor was correct. **(3 marks)**

(d) The man was told by his doctor to continue taking the antibiotics for the full 7 days. Use the information in the graph to explain why it was important that he did not stop taking the antibiotics as soon as he felt better.

(3 marks)

(Total for question 2 = 10 marks)

3 A biotechnology company wanted to develop a new strain of insect-resistant broad beans by inserting the gene for the Bt toxin into the broad bean cells to make 'Bt broad beans'.

(a) The Bt toxin gene was obtained from the bacterium *Bacillus thuringiensis*. State the type of enzyme used to cut the Bt toxin gene out of the *Bacillus thruingiensis* DNA. **(1 mark)**

(b) The table shows the processes involved in preparing the Bt broad bean plants, but they are not in the correct order. Complete the table by putting a number in each box to show the correct order. The first one has been completed for you.

Process	Order
A bacterial plasmid is cut open and mixed with the fragments.	
DNA fragments containing the Bt toxin gene are prepared.	1
DNA ligase joins the sticky ends.	
Agrobacterium tumefaciens is used to insert the recombinant plasmid into the broad bean cells.	
The recombinant plasmid is grown in bacteria to make many copies.	

(2 marks)

(c) Explain the importance of sticky ends and DNA ligase in this process. **(3 marks)**

(d) State **one** advantage and **one** disadvantage of using Bt broad bean plants. **(2 marks)**

(Total for question 3 = 8 marks)

4 The diagram shows a bacterial cell and a plant cell.

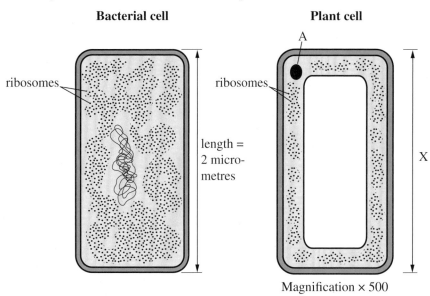

Bacterial cell   **Plant cell**

ribosomes   ribosomes

length = 2 micrometres   X

Magnification × 500

(a) State the name of structure A. **(1 mark)**

(b) (i) Both types of cell contain ribosomes. State the function of a ribosome. **(1 mark)**

(ii) The plant cell contains mitochondria but the bacterial cell does not. State **two** other ways in which the plant cell is different to the bacterial cell. **(2 marks)**

(c) Although the cells are drawn the same size, the magnifications are different. The actual length of the bacterial cell is 2 micrometres. Calculate the actual length, X, of the plant cell in micrometres. Give your answer in standard form and to one decimal place. Show your working. **(3 marks)**

(d) State, with an explanation, the type of microscope that would be used to examine each type of cell. **(2 marks)**

(Total for question 4 = 9 marks)

5 The diagram shows a percentile chart developed by the US Government to monitor the growth of males between the ages of 2 and 20 years. It can be used to monitor both weight and height.

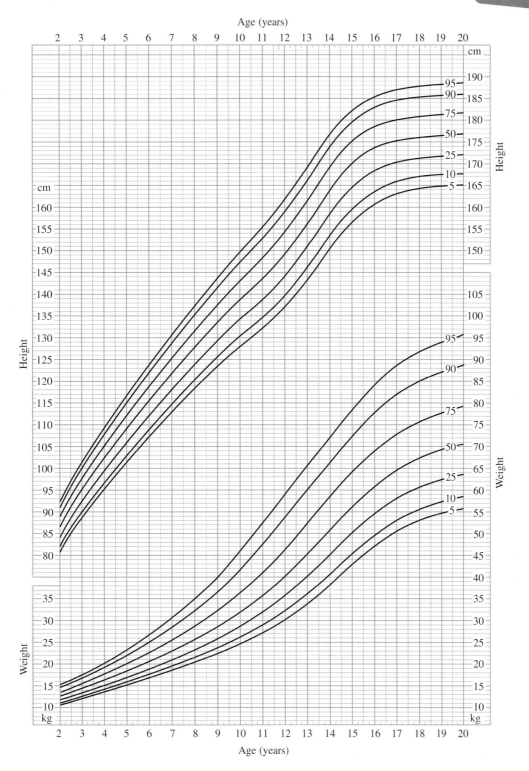

(a) Describe how a doctor or nurse could use this chart to monitor the growth of a boy from the age of 2 to 16 years. **(3 marks)**

(b) The table shows the weight and height records of two boys, A and B, from the ages of 4 to 16.

Age (years)	Height (cm)		Weight (kg)	
	Boy A	Boy B	Boy A	Boy B
4	102	105	18	17
8	127	132	32	27
12	148	155	56	42
16	170	179	83	60

(i) Plot the height and weight of both boys on the percentile chart above. Use '+' for Boy A and 'x' for Boy B. **(4 marks)**

(ii) Explain what your plotted points show about the development of the two boys. **(4 marks)**

(Total for question 5 = 11 marks)   121

6 (a) Identify which of the following statements is correct.

☐ **A** The sequence of bases in a gene represents the phenotype.

☐ **B** The sequence of bases in a gene represents the genotype.

☐ **C** The sequence of amino acids in a gene represents the phenotype.

☐ **D** The sequence of amino acids in a gene represents the genotype.   **(1 mark)**

(b) (i) An allele contains the following DNA sequence ATCGGTCTACCG.

Give the sequence of the mRNA produced from this sequence of DNA.   **(1 mark)**

(ii) This allele was found to consist of 990 bases.

Calculate the number of amino acids in the protein that would be produced.   **(1 mark)**

(c) Two proteins, DAZL and PRDM14, are involved in development of sperm cells. Mutations in these genes have been associated with an increased risk of developing testicular cancer. Almost 100% of all testicular cancers can be completely cured if diagnosed early.

Explain how the Human Genome Project has made it possible to improve early diagnosis of testicular cancer in men with a family history of testicular cancer.   **(3 marks)**

(Total for question 6 = 6 marks)

7 A student was investigating the effect of two different antibiotics on the growth of bacteria. She prepared 10 agar plates containing either antibiotic A or antibiotic B. She then spread bacteria on each plate and incubated them for 3 days. After 3 days she counted the number of bacterial colonies on each plate. Her results are shown in the table.

	Number of bacterial colonies per plate					
	1	**2**	**3**	**4**	**5**	**Mean**
Antibiotic A	5	0	1	3	2	
Antibiotic B	55	50	54	53	56	

(a) (i) State **one** precaution she should take to prevent growth of pathogenic bacteria.   **(1 mark)**

(ii) State **two** precautions she should take to prevent contamination of the plate with other microorganisms.

(2 marks)

(b) (i) Calculate the mean number of colonies for antibiotic A, and for antibiotic B. Write your answers in the table.   **(2 marks)**

(ii) Describe what the results of the experiment show.   **(2 marks)**

(Total for question 7 = 7 marks)

8　The graph shows the results of two studies into the effect of alcohol consumption on the risk of developing liver disease. One group (solid line) consisted only of males and the other group (dotted line) consisted only of females.

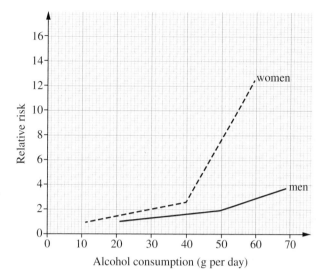

(a)　State and explain the relationship between alcohol consumption and relative risk of liver disease for men.

(3 marks)

(b)　The scientists concluded that alcohol intake was a greater risk for women than for men. Explain how their results supported this conclusion.　**(2 marks)**

(c)　A patient has a height of 1.8 m and a body mass of 100 kg.

What is his BMI?　**(1 mark)**

☐　**A** 30.9

☐　**B** 61.7

☐　**C** 27.8

☐　**D** 55.6

(Total for question 8 = 6 marks)

9　(a)　In humans there are two types of cell division: mitosis and meiosis. The table gives several statements about cell division. Tick one box in each row if the statement is true for mitosis only, for meiosis only or for both mitosis and meiosis. The first row has been completed for you.

Statement	Mitosis only	Meiosis only	Both mitosis and meiosis
used for growth and replacement of cells	✓		
used for production of gametes			
before the parent cell divides each chromosome is copied			
produces genetically identical cells			
halves the chromosome number			

(4 marks)

(b)　There are two types of reproduction: sexual and asexual.

(i)　State **one** advantage and **one** disadvantage for sexual reproduction.　**(2 marks)**

(ii)　State **one** advantage and **one** disadvantage for asexual reproduction.　**(2 marks)**

(c) In sexual reproduction in animals, an egg fuses with a sperm. The diagram below shows the inheritance of X and Y chromosomes.

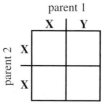

(i) Draw a ring round the part or parts of the diagram that represent(s) sperm cells. **(1 mark)**

(ii) Complete the Punnett square above and use it to calculate the chance of producing female offspring.

(3 marks)

(d)* Discuss the importance of sexual reproduction in the evolution of new species. **(6 marks)**

(Total for question 9 = 18 marks)

10 The diagram shows the neurones and other parts of the body involved in the response to touching a sharp object.

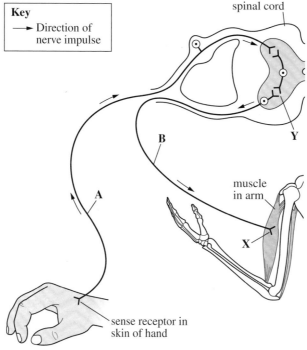

(a) Identify which of the following describes the correct sequence of events following touching a sharp object.

☐ **A** sensory receptor → sensory neurone → motor neurone → relay neurone

☐ **B** sensory receptor → muscle → motor neurone → relay neurone

☐ **C** sensory receptor → relay neurone → sensory neurone → motor neurone

☐ **D** sensory receptor → sensory neurone → relay neurone → motor neurone **(1 mark)**

(b) (i) State the name of the structure labelled Y on the diagram. **(1 mark)**

(ii) Describe the events that occur at point Y that allow the impulse to be passed on from one neurone to the next. **(3 marks)**

(c) (i) Explain what would be the effect of an injury to the neurone labelled A. **(2 marks)**

(ii) Explain how the effect of damage to the spine in a road traffic accident would differ from injury to neurone A. **(3 marks)**

(d) Stem cell therapy has been suggested as a possible treatment for patients with spinal cord injuries.

(i) Describe how a doctor could investigate the extent of the injury to the patient's spinal cord. **(3 marks)**

(ii)*Explain how stem cell therapy could be used in the future to treat spinal cord injury. In your answer you should discuss the sources of stem cells as well as the ethical implications. **(6 marks)**

(Total for question 10 = 19 marks)

Timed Test 2

Time allowed: 1 hour 45 minutes

Total marks: 100

Edexcel publishes official Sample Assessment Material on its website. This practice exam paper has been written to help you practise what you have learned and may not be representative of a real exam paper.

1 (a) The diagram shows the apparatus used to measure the energy content of a piece of food. Once the food was burning, it was moved under the test tube and the temperature rise in the water was measured.

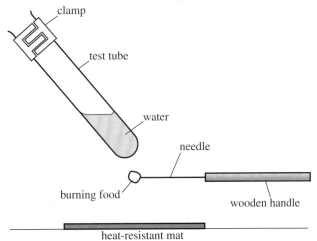

   (i) Identify one piece of equipment missing from the diagram. **(1 mark)**

   (ii) Give **two** reasons why the observed temperature rise might be less than expected. **(2 marks)**

   (iii) A bomb calorimeter is a device that gives an accurate measurement of the energy content of foods. In one experiment, 8.3 g of rice flour is completely burnt in a bomb calorimeter. The temperature of 500 g of water increases by 60.8 °C.

     Calculate the energy content of rice flour in kJ/g. **(3 marks)**

     Use this equation to calculate the energy transferred to heat the water:

     energy transferred (J) = mass of water (g) × 4.2 × change in temperature (°C)

(b) Rice is the main food source for many people. The rice flour tested contained 80% carbohydrate and 6% protein.

   (i) State the name of the test used to show that a food contains protein. **(1 mark)**

   (ii) State the colour you would see with a positive test for protein. **(1 mark)**

(c) Scientists are developing new strains of rice with increased protein content.

   Suggest an explanation for why this is important. **(2 marks)**

(Total for Question 1 = 10 marks)

2 Various hormones and drugs are used in assisted reproductive therapy. This term covers a number of treatments to help couples who are having difficulty conceiving a child. One such hormone is FSH.

(a) (i) State where FSH is produced. **(1 mark)**

   (ii) Describe the effect of FSH on the ovary. **(1 mark)**

(b) Explain how an increase in oestrogen leads to ovulation during a normal menstrual cycle. **(2 marks)**

(c) The drug clomifene can be used to treat women who have difficulty conceiving.

   (i) Explain how clomifene can help such women conceive. **(3 marks)**

   (ii) Explain why clomifene is used to stimulate ovulation in women undergoing IVF, even if they ovulate naturally. **(2 marks)**

(Total for Question 2 = 9 marks)

3 The graph shows the heart rate of an adult male over a 24 hour period.

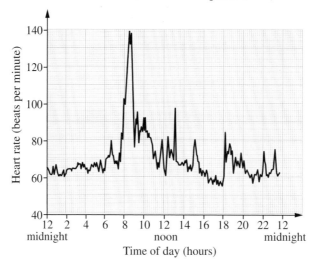

(a) (i)   Use the graph to estimate this person's resting heart rate. **(1 mark)**

  (ii) The man attended a 1 hour spinning (indoor cycling) class during the day.

       Use the graph to estimate the start time of the class. **(1 mark)**

(iii) About an hour before the class started, the man walked uphill to the gym where the class was held. He then rested until the class started.

       Explain how the trace supports this. **(2 marks)**

(iv) During the second half of the class, the man found it harder to maintain the cycling pace.

       Explain why the man found it harder to maintain the cycling pace. **(2 marks)**

  (v) The man noticed that his heart rate remained higher than normal for some time after the end of the class.

       Explain why his heart rate remained high after the class had finished. **(2 marks)**

(b) (i)   The table shows the stroke volume and heart rate for two people measured while they were at rest.

       Complete the table by calculating the cardiac output for each person. Include the units for cardiac output.

	Stroke volume (cm^3)	Heart rate (beats per minute)	Cardiac output	Units
Person A	95	52		
Person B	58	72		

(3 marks)

  (ii) One of these was a trained athlete, the other was untrained.

       Explain which is the trained athlete. **(2 marks)**

(Total for Question 3 = 13 marks)

4 The diagram shows the part of the lung where gas exchange takes place.

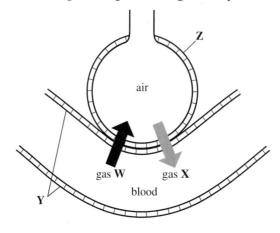

(a) (i) State the names of the structures labelled **Y** and **Z** in the diagram. **(2 marks)**

   (ii) State which process is used to transport gases **W** and **X** in the directions shown. **(1 mark)**

   (iii) Give the name of gas **W**. **(1 mark)**

   (iv) Describe how gas **X** is carried in the blood. **(1 mark)**

(b) (i) A scientist recently estimated that the average human lung has 480 million alveoli and that 1 million alveoli have a surface area of 0.15 m^2.

   Calculate the total surface area of a human lung. **(1 mark)**

   (ii) An athlete trains to run a marathon. The surface area of the athlete's lungs increases to 82 m^2.

   Use Fick's law to explain how this will improve the athlete's performance. **(3 marks)**

(Total for Question 4 = 9 marks)

5 An epiphyte is a plant that grows harmlessly on another species of plant. The epiphyte obtains its water and nutrients from air, rain and debris that accumulate around it.

(a) (i) Mistletoe is a parasitic plant. Explain how a parasite differs from an epiphyte. **(3 marks)**

   (ii) Nitrogen-fixing bacteria grow in the root nodules of legumes.

   Explain why nitrogen-fixing bacteria are mutualists rather than parasites. **(3 marks)**

(b) Parasitic and epiphytic plants have different adaptations to their environment.

   (i) Describe **one** adaptation an epiphyte growing in a tropical rainforest might have. **(1 mark)**

   (ii) Some epiphytes grow in places where there is little water.

   Describe **two** adaptations these plants might have to help them survive in these conditions. **(2 marks)**

(Total for Question 5 = 9 marks)

6 (a) Adrenalin and noradrenalin are chemically similar molecules. However, adrenalin is a hormone and noradrenalin is a neurotransmitter.

   (i) State the organ where adrenalin is produced. **(1 mark)**

   (ii) Describe how adrenalin reaches its target organs. **(2 marks)**

   (iii) Describe **two** ways in which hormonal communication is different to nervous communication.

(2 marks)

(b) Adrenalin causes muscle cells to convert glycogen to glucose.

   (i) State the name of one other hormone that has the same effect. **(1 mark)**

   (ii) State the name of the disease caused by the body not being able to regulate the concentration of glucose in the blood. **(1 mark)**

(c)* Homeostasis is the term used to describe how the internal environment of the body is kept constant and involves the process of negative feedback. Thermoregulation describes how the temperature of the body is maintained within a narrow range.

   Explain how negative feedback operates in thermoregulation. **(6 marks)**

(Total for Question 6 = 13 marks)

7 A group of students were undertaking a survey of an area of land alongside a path that crossed a field and entered a piece of woodland.

(a) (i) State **two** abiotic factors that might influence the distribution of plant species in the woodland.

(2 marks)

   (ii) State **two** biotic factors that might influence the distribution of plant species next to the path in the field. **(2 marks)**

(b) Describe how the students should survey the abundance of different plant species growing alongside the path from the field and into the wood. **(4 marks)**

(Total for Question 7 = 8 marks)

8   A student carried out an investigation into osmosis in potato pieces. The student cut five pieces of potato, weighed them and then placed them into different concentrations of sucrose solution. After one hour the student removed the potato pieces from the sucrose solution and weighed them again. The student's results are shown in the table.

Concentration of sucrose solution (mol dm^{-3})	Initial mass of potato (g)	Final mass of potato (g)	Percentage increase / decrease
0.0	5.2	5.4	
0.5	5.6	5.6	
1.0	5.6	5.4	
2.0	5.0	4.6	
3.0	5.2	4.2	

(a) (i)   Complete the table by calculating the percentage increase or decrease in weight of the potato pieces. Give your answers to one decimal place.   **(3 marks)**

   (ii)   State **one** variable the student would need to control during the experiment.   **(1 mark)**

   (iii)   Describe **one** way in which the student could improve the experiment.   **(1 mark)**

   (iv)   Use the results to estimate the solute concentration of the potato cells.

          Explain your answer.   **(2 marks)**

(b) People with kidney failure can be treated with dialysis.

    Describe how dialysis works to make sure that the patient's blood has the right concentration of substances.   **(4 marks)**

                                                                            (Total for Question 8 = 11 marks)

9   (a) The diagram shows a specialised type of plant tissue.

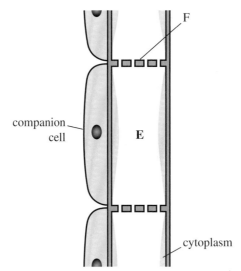

   (i)   Which specialised plant tissue is shown in the diagram?

         ☐ **A** xylem   ☐ **C** mesophyll

         ☐ **B** phloem   ☐ **D** root hair   **(1 mark)**

   (ii)   Identify the parts labelled E and F.   **(2 marks)**

   (iii)   Explain why the companion cell contains large numbers of mitochondria.   **(3 marks)**

(b) (i)   Temperature can be a limiting factor in photosynthesis. State **one** other factor that can limit the rate of photosynthesis.   **(1 mark)**

   (ii)   The graph shows how the rate of photosynthesis changes as temperature is increased.

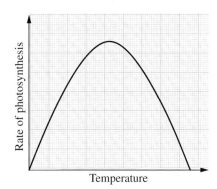

Explain why the rate of photosynthesis changes as the temperature is increased.   **(3 marks)**

(Total for Question 9 = 10 marks)

10 (a) Indicator species are used to assess levels of pollution.

Which of the following best describes an indicator species?

☐ **A** a species that can tolerate high levels of pollution

☐ **B** a species that can only live where there is no pollution

☐ **C** a species where the individuals change colour where pollution is present

☐ **D** a species that lives only in a small number of habitats   **(1 mark)**

(b) Sea lice infest wild and farmed salmon. The graph shows the mean number of sea lice per salmon in farmed salmon in Ireland. New control measures were introduced in 2007.

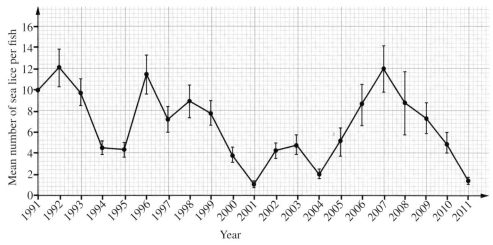

Calculate the percentage increase in the mean number of sea lice per salmon between 2004 and 2007.

(1 mark)

(c)* Killary Harbour is a 16 kilometre fjord on the west coast of Ireland where two types of aquaculture are carried out – salmon farming and mussel farming. Mussel farming involves suspending ropes in the water and mussel 'seed' (immature shellfish) are attached. The mussels then grow naturally over a period of 12–15 months. Mussels feed on microscopic organisms that they filter from the sea water. This can remove phosphates and nitrates from the water. Some people are opposed to salmon farming because they believe it is harmful to the environment. In contrast with salmon farming, mussel farming is thought to have less impact on the environment.

Evaluate the two types of aquaculture in terms of their environmental impact.   **(6 marks)**

(Total for Question 10 = 8 marks)

Answers

1. Plant and animal cells

1 (a) B (**1**)

(b) C (**1**)

2 (a) Muscles need large amounts of energy for contraction (**1**). This energy is supplied from respiration in mitochondria (**1**).

(b) Mitochondria release energy and all cells need energy (**1**), but only leaf (and stem) cells are exposed to light and so have chloroplasts for photosynthesis (**1**).

3 Cell membrane controls what enters and leaves the cell (**1**); cell wall helps to support the cell / helps it keep its shape (**1**).

4 Enzymes are proteins and proteins are made on ribosomes (**1**); fat cells don't produce as many proteins as pancreatic exocrine cells (**1**).

2. Different kinds of cell

1 C (**1**)

2 (a) A = acrosome (**1**); B = flagellum (**1**)

(b) A contains enzymes to digest a way through the egg cell membrane (**1**); B is used to move the bacterium towards a food source (**1**).

3 Epithelial cells line tubes (such as trachea) (**1**). Mucus traps dirt / dust / bacteria (**1**) and cilia move mucus along the tubes away from the lungs (**1**).

4 The egg cell contains nutrients in the cytoplasm (**1**) to supply the growing embryo (**1**).

3. Microscopes and magnification

1 Light microscopes magnify less than electron microscopes. (**1**) The level of cell detail seen with the electron microscope is greater (**1**) because it has a higher resolution (**1**).

2 (a) because it has a nucleus (**1**) and eukaryotic cells have nuclei (**1**)

(b) (i) $(23 / 5) \times 2$ (**1**) = 9.2 µm (**1**)

(ii) $(4 / 5) \times 2$ (**1**) = 1.6 µm (**1**)

(c) Nuclei are large enough to be seen with a light microscope (**1**) but mitochondria are too small and can only be seen with an electron microscope (**1**) because it has a higher resolution / greater magnification (**1**).

3 (a) light microscope: 2.5 µm × 1000 (**1**) = 2500 µm (or 2.5 mm) (**1**); electron microscope: 2.5 µm × 100 000 = 250 000 µm (or 250 mm or 25 cm or 0.25 m) (**1**) (Note that you get the mark for correct use of the formula just once even though you use it twice.)

(b) The electron microscope (**1**) because it would show more detail / has the correct resolution (**1**).

4. Dealing with numbers

1 picometre, nanometre, micrometre, millimetre, metre (**1**)

2 5 picometres (**1**), 0.25 grams (**1**), 0.00025 kilograms (**1**), 2500 millimetres (**1**)

3 true (**1**), false (**1**), false (**1**), true (**1**)

4 (a) 0.0309 / 1 000 000 (**1**) = 3.1×10^{-8} m (**1**)

(b) 0.163 / 250 000 (**1**) = 6.5×10^{-7} m (**1**)

(c) 0.0078 / 800 (**1**) = 9.8×10^{-6} m (**1**)

5. Using a light microscope

1 (a) (i) to reflect light through the slide (**1**)

(ii) to hold the slide in place (**1**)

(iii) to move the objective up and down a long way (**1**)

(b) (i) because it could crash into the slide (**1**)

(ii) because it could permanently damage eyesight (**1**)

(c) (i) a desk / bench / built-in lamp (**1**)

(ii) Two from: always start with the lowest power objective under the eyepiece (**1**); clip the slide securely on the stage (**1**); move the slide so the cell you need is in the middle of the (low power) view (**1**); use only the fine focusing wheel with the high power objective (**1**)

2 Three from: go back to using the low power objective (**1**); find the part you need and bring it back to the centre view (**1**); focus on it with the coarse focus (**1**); return to the high power objective (**1**) and use the fine focus wheel to bring the part into focus (**1**).

6. Drawing labelled diagrams

1 (a) Three from: the drawing is in pen rather than pencil (**1**); the title is incomplete (**1**); the magnification is not given (**1**); label lines are not drawn with a ruler (**1**) and cross each other (**1**); not enough cells are shown (**1**) and they are not drawn to scale (**1**); shading should not be used (**1**); lines have been crossed out rather than rubbed out (**1**) and are ragged rather than clear (**1**); the cell membrane can't be seen with the light microscope (**1**)

(b) Clear drawing of all / most of the cells (**1**); cells not of interest drawn just as outlines (**1**); detail of representative sample of cells (**1**); and avoidance of mistakes from 1 (a) (**1**)

2 Width of image = 45 mm (**1**) so magnification = 45 / 0.113 (**1**) = ×398 (or 400) (**1**)

7. Enzymes

1 The shape of the active site of invertase matches the shape of sucrose but not lactose (**1**), so invertase cannot combine with lactose and catalyse its digestion (**1**).

2 (a) The rate rises gradually at first (**1**) then reaches a peak at about 40 °C (**1**) and then drops rapidly (**1**).

(b) (i) At lower temperatures the molecules move slowly (**1**) so substrate molecules take longer to fit into the active site and react (**1**).

(ii) At the optimum temperature (**1**) the enzyme is working at its fastest rate (**1**).

(iii) Two from: higher temperatures cause the active site to change shape (**1**) so it can't hold the substrate as tightly (**1**) / the active site breaks up (**1**) and the enzyme is denatured (**1**).

3 Amylase digests starch in the mouth but is denatured in the stomach (**1**); pepsin has a pH optimum ~2 so digests proteins in the stomach (**1**); pancreatic juice neutralises stomach acidity (**1**); so trypsin and amylase will work in the small intestine (**1**).

8. pH and enzyme activity

1 (a)

pH	2	4	6	8	10
Time (min)	> 10	7.5	3.6	1.2	8.3
Rate (min)	0	0.13	**0.28**	**0.83**	**0.12**

(**2 marks** for all 5 correct, **1 mark** for 3 correct)

(b)

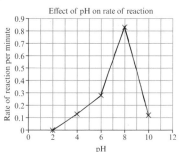

correctly labelled axes (**1**), appropriate use of paper (plotted points filling >50% of space) (**1**), correctly plotted points (**1**), line of best fit (can be smooth curve) (**1**)

(c) Two from: use a water bath to control temperature (**1**); repeat several times and take a mean (**1**); use a more accurate method to determine if the film is clear (**1**); use more intermediate pH values (**1**)

9. The importance of enzymes

1.

Enzyme	Digests	Product(s)
amylase	starch	**sugars / maltose**
lipase	**lipids**	**fatty acids and glycerol**
protease	**proteins**	amino acids

1 mark for each correct row.

2 (a) Many different enzymes are needed because they are specific for different food molecules (**1**); digestion breaks down the food molecules into molecules small enough to be absorbed (**1**).

(b) Synthesis reactions occur too slowly (**1**); enzymes are biological catalysts and speed up reactions (**1**).

3 (a) protease (**1**); needed to break down egg stain, which is made from protein (**1**)

(b) The enzyme is denatured / active site destroyed (**1**) at higher temperatures (**1**), so it would not digest stains as well / would be less active (**1**).

10. Using reagents in food tests

1 (a) (i) wear eye protection / lab coat / gloves (**1**)

(ii) iodine can stain the skin / clothing (**1**)

(b) Mix food with ethanol and shake (**1**); pour some mixture into water and shake again (**1**); cloudy emulsion will form if lipids present (**1**).

2 (a) The level of starch decreases (**1**) and reducing sugars appear (**1**); proteins are present before and after germination (**1**).

(b) During germination enzymes / amylase digest starch (**1**) and produce reducing sugars / maltose (**1**).

11. Using calorimetry

1 (a) Energy released by the combustion of food (**1**) is transferred to the water causing heating, which is measured by a thermometer (**1**).

(b)

	Brown bread	Biscuit	Dried Apricot	Crisps
Energy value of food sample burnt (**J**)	2940	2016	2772	2184
Energy value of food (**kJ per 100 g**)	118	101	55	109

(Correct working (**1**), one row correct (**1**), second row correct (**1**))

(c) (i) The student's results are (a lot) lower. (**1**).

(ii) Two from: the food may not have burnt completely (**1**); not all the energy is transferred to the water / energy transferred to the surroundings (**1**); the food was not weighed correctly (**1**)

12. Getting in and out of cells

1 Movement of particles (**1**) from high concentration to low concentration / down a concentration gradient (**1**).

2 Both occur across a partially permeable membrane / involve movement of molecules (**1**).

Active transport requires energy / moves molecules against a concentration gradient / movement from low to high concentration whereas diffusion is passive (**1**).

3 (a) Osmosis is the net movement of water molecules (**1**) across a partially permeable membrane (**1**) from a low solute concentration (**1**) to a high solute concentration (**1**).

(b) Gas particles move down the concentration gradient (**1**); there is high concentration of oxygen in the air, low concentration in blood (**1**) and low concentration of carbon dioxide in the air, high concentration in the blood (**1**).

(c) Glucose must be moved against a concentration gradient (**1**) by active transport that requires energy (**1**).

13. Osmosis in potatoes

1 Four from: Cut pieces of potato, making sure size / length is the same (**1**); measure mass (**1**); leave in solution for 20 minutes / same time (**1**). Remove from the solution, then measure mass again (**1**). Blot dry before each weighing (**1**).

2 (a)

Sucrose concentration (mol per dm³)	Initial mass (g)	Final mass (g)	Change in mass (g)	Percentage change (%)
0.0	19.15	21.60	2.45	12.8
0.1	18.30	19.25	0.95	5.2
0.2	15.32	14.85	−0.47	−3.1
0.3	16.30	14.40	−1.90	−11.7
0.5	18.25	16.00	−2.25	−12.3
1.0	19.50	17.20	−2.30	−11.8

Use of correct method (change in mass / initial mass) × 100 (**1**), all six correct (**1**).

(b) (i)

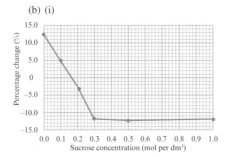

correctly labelled axes (**1**), correctly plotted points (**1**), line of best fit (**1**)

(ii) Solute concentration in range 0.17 to 0.18 mol per dm³ (**1**)

14. Extended response – Key concepts

*Answer could include the following points:

- The dye could enter by diffusion or by active transport.

- Active transport acts against a concentration gradient.

- Removal of all the dye from the solutions shows it is working against a concentration gradient, otherwise the solution would still contain some dye.

- Active transport involves enzymes.

- Enzymes work more slowly at lower temperatures.

- Enzymes are destroyed / denatured at high temperature.

- Further experiments: observe cells under light microscope to see if the dye is inside the cells, repeat experiment at different pH values.

15. Mitosis

1 (a) B (**1**)

(b) interphase, prophase, metaphase, anaphase, telophase (**1**)

2 To start with there is 1 cell; after 1 hour this divides into 2 cells. After 2 hours 4 cells. After 3 hours 8 cells. After 4 hours 16 cells (**1**).

3 (a) A = anaphase (**1**); B = metaphase (**1**)

(b) A = because chromatids are being pulled to each pole (**1**); B = because chromosomes are lined up along the middle of the cell (**1**)

(c) The nuclei divide by mitosis (**1**) but the two new cells do not separate from one another (**1**).

16. Cell growth and differentiation

1 (a) zygote (**1**)

(b) mitosis (**1**)

2 (a) meristem / root tip / shoot tip (**1**)

(b) Vacuoles take in water by osmosis (**1**) and this causes the cell to elongate (**1**).

3 (a)

Type of specialised cell	Animal or plant
sperm	animal
xylem	**plant**
ciliated cell	**animal**
root hair cell	**plant**
egg cell	**animal**

3 marks for 4 or 5 correct, **2 marks** for 3 correct, **1 mark** for 2 correct.

(b) Plants: mesophyll cell / guard cell / phloem (**1**). Animals: small intestine cell / hepatocyte / red blood cell / nerve cell / bone cell / (smooth) muscle cell (**1**).

4 (a) become specialised (**1**) to perform a particular function (**1**)

(b) because there are many different kinds of specialised cells (**1**) that can carry out different processes more effectively (**1**)

17. Growth and percentile charts

1 (a) C (**1**)

(b) 47.5 − 46.0 (**1**) = 1.5 cm (±0.2 cm) (**1**)

2 (a) 15.35 − 12.75 = 2.60 g (**1**); (2.60 / 12.75) × 100 = 20.4% (**1**)

(b) Any suitable, such as: height (**1**), measured with a ruler ensuring the stem is vertical (**1**); shoots / leaves (**1**) by counting number (**1**)

18. Stem cells

1. (a) All the cells in an embryo are stem cells, but in an adult stem cells are found only in some tissues such as bone marrow (**1**). Embryonic stem cells can differentiate into many cell types / all diploid cells, but adult stem cells can only differentiate into one type of cell / limited types of cell (**1**).

   (b) (i) meristem (**1**)

   (ii) tips of root (**1**) and tips of shoot (**1**)

2. (a) to replace damaged / worn out cells (**1**)

   (b) Differentiated cells cannot divide / embryonic stem cells can divide, to produce other kinds of cell (**1**).

3. (a) Embryonic stem cells could be stimulated to produce nerve cells (**1**) then transplanted into the patient's brain (**1**).

   (b) (i) advantage: easy to extract / can differentiate into nerve cells (**1**); disadvantage: requires destruction of embryo / may be rejected / may cause cancer (**1**)

   (ii) advantage: does not destroy embryos / will not be rejected (**1**); disadvantage: may cause cancer / may not differentiate into nerve cells (**1**)

19. The brain and spinal cord

1. (a) brain **and** spinal cord (**1**)

   (b) (i) regulates heartbeat / breathing (**1**)

   (ii) coordinates and controls precise and smooth movement (**1**)

   (c) Two from: control voluntary movement (**1**); interpret sensory information (**1**); responsible for learning (**1**) and memory (**1**)

2. Her running is coordinated by the cerebellum, which controls smooth movement and keeps her balanced (**1**). The cerebral hemispheres (**1**) interpret the sensory information from her ears while listening to music and also from her eyes when she sees her friend. Her heart rate and breathing rate are controlled by the medulla oblongata (**1**). Waving to her friend is controlled by the cerebral hemispheres (**1**).

3. Four from: His spinal cord was broken above the waist (**1**) but below the neck (**1**) so nerve impulses could no longer pass to the parts of the body below the break (**1**) and he would be unable to control or move parts of the body below the break (**1**). The paralysis is permanent because neurones cannot divide to replace damaged cells (**1**).

20. Treating damage and disease in the nervous system

1. (a) The skull protects the brain (**1**).

   (b) The spine protects the spinal cord (**1**).

2. (a) It is difficult to get medicines into the central nervous system (**1**) because of the bony skull and spine / because they cannot cross the blood–brain barrier (**1**).

   (b) (i) The bony skull makes surgery difficult (**1**) and there is a risk of damage to healthy tissue (**1**).

   (ii) because it could cause damage to other parts of the brain as it grows (**1**)

   (iii) Damage to nerve cells could cause paralysis (**1**) and nerve cells can't regrow so it would be permanent (**1**).

3. (a) to make it easier to remove the tumour (**1**) without removing too much healthy tissue (**1**)

   (b) because some cancer cells might be left behind (**1**) and these will be killed by radiotherapy (**1**)

21. Neurones

1. (a) Dendrons carry electrical impulses towards the cell body (**1**).

   (b) Axons carry electrical impulses away from the cell body (**1**).

2. A, axon endings; B, axon; C, cell body; D, dendron; E, myelin sheath; F, receptor cells (in skin) (all correct, **3 marks**; 4 or 5 correct, **2 marks**; 2 or 3 correct, **1 mark**)

3. (a) The cell body of a sensory neurone is in the middle of the axon (**1**). The cell body of a motor neurone is at the beginning of the neurone (**1**).

   (b) Three from: Dendrites collect impulse from central nervous system (**1**); axon carries this over long distances through the body (**1**); nerve ending transmits the impulse to an effector (**1**); myelin sheath speeds up transmission (**1**).

4. (a) Myelin sheath speeds up transmission (**1**) because the impulse jumps from one gap to another (**1**).

   (b) Their movement would be impaired / made difficult (**1**) because the nerve impulses to muscles would be slower (**1**).

22. Responding to stimuli

1. (a) synapse (**1**)

   (b) Neurone Y (**1**); because it is carrying impulses to an effector / muscle (**1**)

   (c) When an electrical impulse reaches the end of neurone X it causes the release of neurotransmitter (**1**) into the gap between the neurones. This substance diffuses (**1**) across the synapse / gap (**1**) and causes neurone Y to generate an electrical impulse (**1**).

2. (a) Three from: stimulus is detected by receptors (**1**); a nerve impulse travels along a sensory neurone (**1**) then through a relay neurone in the brain / CNS / spinal cord (**1**) and along a motor neurone to an effector (**1**).

   (b) light / movement (**1**) because it causes the eyelid to blink (**1**)

3. Reflex responses are automatic / very fast (**1**) so they protect the body / help avoid danger (**1**) which increases chances of survival (**1**).

23. The eye

1. (a) A, cornea; B, pupil; C, lens; D, iris; E, ciliary muscles; F, retina (**3 marks** for all correct, **2 marks** for 4 or 5 correct, **1 mark** for 2 or 3 correct)

   (b) Both are transparent to let light through (**1**); both are curved to refract / bend / focus light (**1**); the shape of structure C can change / become more or less curved (**1**).

2. The iris changes its size by muscle contraction and relaxation (**1**). It does this to control how much light enters (**1**).

3. (a) To form an image the light rays must converge / be focused / be refracted (**1**) onto the retina (**1**). This occurs as light passes from air through the curved surface of the cornea and lens (**1**).

   (b) (The ciliary muscles contract) making the lens fatter (**1**), so increasing the refraction / causing the light to bend or converge more (**1**).

4. They need to be able to see in very dim light (**1**) and rod cells are sensitive at low light intensity (**1**), cones only work in bright light (**1**).

24. Eye problems

1. (a) short sight (**1**)

   (b) The lens makes the light rays diverge (**1**), so that the image moves away from the lens / closer to the retina (**1**).

   (c) (i) must be clean / free of microbes to avoid infections (**1**)

   (ii) The cells in the cornea need oxygen for respiration / to stay alive (**1**).

2. Because the lens is cloudy so light is scattered (**1**) and doesn't focus on the retina (**1**).

3. As they get older the lens does not bend enough (**1**) and so they cannot focus on close objects (**1**).

4. (a) Cone(s) (**1**)

   (b) Either red or green cones are missing (**1**) so the person cannot distinguish between red and green (**1**).

25. Extended response – Cells and control

*Answer could include the following points:

- Stages of mitosis described as part of the cell cycle.
- Production of genetically identical daughter cells.
- Diploid number maintained in all cells except gametes.
- Involves replication of DNA.
- Description of cell differentiation.
- Examples of specialised cell types.
- Importance of stem cells: in embryo to produce all different kinds of cell in the body; in adult for growth and repair.

26. Asexual and sexual reproduction

1.

Feature	Sexual reproduction	Asexual reproduction
need to find a mate	needs to find a mate	no need to find a mate
mixing of genetic information	mixes genetic information from each parent	no mixing of genetic information
characteristics of offspring	offspring show variety of characteristics	offspring have same characteristics as parent / each other

1 mark for each correct row.

2. (a) runners: asexual (**1**) because only one parent / offspring all identical to parent

(1); fruits: sexual **(1)** because there are two parents / fusion of gametes **(1)**

(b) asexual: plant can make use of beneficial conditions because more plants produced quickly **(1)**; sexual: offspring may be better adapted if conditions change because of variation **(1)** OR prevents overcrowding because seeds spread more widely **(1)**

(c) asexual: no genetic variation so plants may die / not grow as well if conditions change **(1)**; sexual: requires two parents (pollination) / requires energy to produce fruits **(1)**

3 Two from: energy needed to find a mate **(1)**, in courtship behaviour **(1)**, in gamete production **(1)**

27. Meiosis

1 (a) (i) half the number of chromosomes / one set of chromosomes **(1)**

    (ii) sex cells **(1)**

(b) male: sperm **(1)**; female: egg / ovum **(1)**

2 (a) 10 **(1)**

(b) Each daughter cell has only half of chromosomes / genes / DNA from each parent **(1)** so different daughter cells have different combinations of chromosomes / genes / DNA **(1)**

3 (a) DNA replication **(1)**

(b)

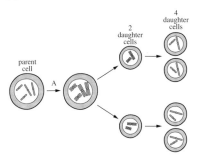

2 daughter cells, then 4 daughter cells **(1)**, one of each pair consisting of duplicated chromosomes in 2 daughter cells **(1)**, 4 daughter cells contain one copy of each pair **(1)**

4 Mitosis maintains the diploid number **(1)** and produces cells that are identical to the parent **(1)**. It is used for growth **(1)**. Meiosis creates gametes that have half the number of chromosomes **(1)**. Fertilisation restores the diploid number **(1)**.

28. DNA

1 (a) genome **(1)**

(b) A chromosome consists of a long molecule of DNA packed with proteins **(1)**; a gene is a section of DNA molecule / section of chromosome that codes for a specific protein **(1)**; DNA is the molecule containing genetic information that forms part of chromosomes **(1)**.

2 (a) double helix **(1)**

(b) (i) 4 **(1)**

    (ii) weak hydrogen bonds between complementary bases **(1)**

3 (a) The structure consists of repeated nucleotides / monomers **(1)**.

(b) A, base **(1)**; B, sugar / ribose **(1)**; C, phosphate **(1)**

4 (a) TACCCG **(1)**

(b) because there are complementary base pairs **(1)**; A always pairs with T, C with G **(1)**

29. Protein synthesis

1 RNA polymerase binds to a non-coding region of DNA in front of a gene **(1)** and then moves along the template strand **(1)** adding complementary RNA nucleotides **(1)**; these then link together to form mRNA **(1)**.

2 (a) ribosomes **(1)**

(b) travels out of nucleus through small holes in the membrane / nuclear pores **(1)**

3 (a) (i) 153 **(1)**

    (ii) $3 \times 153 = 459$ **(1)**

(b) Four from: ribosome moves along the mRNA reading one triplet codon at a time **(1)**; tRNA brings amino acids to the ribosome **(1)**; complementary bases on tRNA pair with bases on mRNA **(1)**; ribosome moves along joining the amino acids to make a polypeptide chain **(1)**, which then folds up to form the myoglobin protein **(1)**.

30. Gregor Mendel

1 (a) Red hair is either present or absent **(1)** and a parent with red hair may not have red-haired children **(1)**.

(b) These could not be caused by blending **(1)** because the characteristics were either present or absent **(1)**.

2 (a) so that he would know what factors they had **(1)** because they always produce identical offspring when crossed with a pea of the same type **(1)**

(b) so that he knew which plants had been crossed **(1)** because this excludes pollen from other plants **(1)**

(c) that the factor for yellow seeds was dominant to the factor for green seeds **(1)** and the factor for round seeds was dominant to the factor for wrinkled seeds **(1)** because although there were no green or wrinkled seeds after the first cross, some appeared in the second generation **(1)**

(d) yellow round seeds **(1)**

31. Genetic terms

1 (a) (i) different forms of the same gene that produce different variations of the characteristic **(1)**

    (ii) Genotype shows the alleles that are present in the individual, e.g. Bb or BB **(1)**, whereas phenotype means the characteristics that are produced, e.g. brown eyes or blue eyes **(1)**.

(b) bb **(1)**, BB **(1)**, Bb **(1)**

(c) bb **(1)** because to have blue eyes she must have two recessive alleles **(1)**

2 There are two copies of each chromosome in body cells **(1)**; each copy has the same genes in the same order **(1)**; a gene is a short piece of DNA at a point on a chromosome **(1)**; genes come in different forms called alleles that produce different variations of the characteristic **(1)**.

32. Monohybrid inheritance

1 (a) correct gametes **(1)**; correct genotypes **(1)**

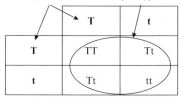

(b) 25% of the offspring from this cross will be short. I know this because tt is short **(1)** and one in four of the possible offspring are tt **(1)**.

(c) $\frac{3}{4}$ / 75% **(1)**

2 correct parent genotypes Gg **(1)** and gg **(1)**; correct gametes G, g, g, and g **(1)**; correct offspring genotype Gg, gg, Gg, gg **(1)**

33. Family pedigrees

1 C **(1)**

2 (a) two **(1)**

(b) one **(1)**

(c) Person 4 does not have cystic fibrosis. This means that they must have one dominant allele from their father **(1)**. But they must have inherited a recessive allele from their mother **(1)**. This means that their genotype is Ff **(1)**.

(d) Two healthy parents (person 3 & person 4) **(1)** produce a child (person 8) with CF **(1)**.

34. Sex determination

1 (a) X **(1)**; the girl has two X chromosomes, one from each parent **(1)**

(b) (i) **1 mark** for parental sex chromosomes and **1 mark** for all possible children's chromosomes

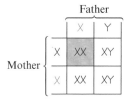

    (ii) female **(1)**

2 (a) 50% / ½ / 0.5 **(1)**; depends on which sperm fertilises the egg **(1)** as half the sperm will carry a male sex chromosome / Y chromosome and half the sperm will carry a female sex chromosome / X chromosome **(1)**.

(b) The statement is not correct **(1)**; the probability of having a child who is a boy is always 50% **(1)**.

35. Inherited characteristics

1 This is when a heterozygous individual shows the effect of both alleles. **(1)**

2 (a) A = $I^A I^A$ **(1)**; AB = $I^A I^B$ **(1)**

(b) group AB, I^AI^B (**1**); because some children received the I^B allele from the mother (**1**) and others received the I^A allele (**1**)

(c) A child who is group B must receive an I^B allele from each parent (**1**), but the man only has I^A alleles so cannot be the father (**1**).

3   Correct Punnett square and gametes (**1**); correct genotypes (**1**); 25% group A, 50% group AB, 25% group B (**1**)

		Father	
		I^A	I^B
Mother	I^A	I^AI^A Group A	I^AI^B Group AB
	I^B	I^AI^B Group AB	I^BI^B Group B

36.   Variation and mutation

1   (a) Students in a year 7 class will show differences in mass caused by genetic variation (**1**) as well as environmental variation (**1**).

(b) Identical twins will only show differences caused by environmental variation (**1**).

2   (a) Mean height = (181 + 184 + 178 + 190 + 193 + 179) / 6 (**1**) = 184.2 cm (**1**)

(b) Four from: height is determined partly by genetic factors (**1**) and partly by environmental factors (**1**) such as nutrition (**1**); different children have inherited different alleles from their parents (**1**); parents have different heights so will pass on different alleles for height (**1**); the mean height of the children is greater than that of the parents because of better nutrition / they take after their father more than their mother (**1**)

3   (a) Two from: no effect (**1**) small effect (**1**) significant effect (**1**) on phenotype

(b) A change in the base sequence of DNA (**1**) leads to a large change in the amino acid sequence of the protein (**1**) so that it has a different structure to the normal protein (**1**).

37.   The Human Genome Project

1   (a) the sequence of bases on all human chromosomes (**1**)

(b) Advantages, two from: a person at risk from a genetic condition will be alerted (**1**); distinguishing between different forms of disease (**1**); tailoring treatments for some diseases to the individual (**1**)

Disadvantages: people at risk of some diseases may have to pay more for life insurance (**1**); it may not be helpful to tell someone they are at risk from an incurable disease (**1**)

2   Advantages, any two from: she could have earlier / more frequent screening for breast cancer (**1**), she could consider surgery to remove the breast / mastectomy (**1**), her doctor might prescribe drugs to reduce the risk of developing cancer (**1**); disadvantages, any two from: it might make her more worried / anxious (**1**), just because she has the mutation doesn't mean she will develop breast cancer (**1**), could have unnecessary surgery / medication (**1**).

38.   Extended response – Genetics

*Answer could include the following points:

- The gene coding for p53 in the DNA; found in the nucleus; unzips; used as a template; to make mRNA; by complementary base pairing; this is transcription.
- mRNA contains uracil in place of thymine; mRNA travels through the cytoplasm; to the ribosome; ribosome reads the mRNA; in units of three bases known as a codon / triplet; this is called translation.
- Each triplet codes for an amino acid; carried to the ribosome by a tRNA molecule; which uses complementary base pairing.
- Amino acid adds to form a chain called a polypeptide; continues until all 393 amino acids to make p53 are linked together.
- Some mutations will not affect the p53 protein, others may increase the risk of cancer.
- The Human Genome Project makes it possible to screen people for differences in p53; these can be linked to increased risk of cancer.

39.   Evolution

1   (a) any two from: theory involved natural selection (**1**) theory was based on their own work / work of previous scientists (**1**) they presented their work jointly in 1858 (**1**)

(b) Two from: It helps us understand the relationships between different species of organisms (**1**); it explains how new species evolve (**1**); it explains how different species adapt to changes in their environment (**1**).

2   There is variation within a species (**1**); members of the species that are most adapted will survive / those that are less well adapted die (**1**).

3   It will help to classify the new species (**1**) and to find out which other organisms the new species is related to (**1**).

4   (a) There is variation in the amount of antibiotic resistance in a population of bacteria (**1**); the most resistant take the longest to die (**1**), so stopping early means the most resistant will survive and reproduce (**1**) so that all the new population of bacteria will be resistant (**1**).

(b) Use of antibiotics is a form of natural selection (**1**), where only the bacteria with advantageous variation / antibiotic resistance (**1**) will survive and pass these genes on to the next generation (**1**).

40.   Human evolution

1   Three from: toe arrangement (**1**), length of arms (**1**), brain size (**1**), skull shape (**1**).

2   (a) Two from: The older the species, the smaller its brain volume (**1**); negative correlation / as years before present became less, brain volume increased OR positive correlation – as time 'increases' brain volume increases (**1**); greatest increase in brain volume between 2.4 and 1.8 million years ago (**1**); increase in brain volume not linear, increased by 500 cm^3 in 2.6 million years (**1**)

(b) an increase in brain volume / size (**1**) to at least 550 cm^3 (**1**)

3   (a) The ages of the rock layers where the tool was found can be dated (**1**) by measuring the amount of radiation in the layers (**1**).

(b) Three from: smooth area in palm of hand (**1**), will not cut / damage hand (**1**); chipped section away from hand (**1**) (as it) has sharp edges (**1**) for cutting / unlike smooth area (**1**).

41.   Classification

1   Both have limbs with five fingers (**1**) that have evolved / become adapted to different uses (**1**).

2   Plant are autotrophic feeders, while animals are heterotrophic feeders (**1**). Plant cells have cell walls, but animal cells do not have cell walls (**1**). *You could also say*: Plant cells contain chlorophyll, but animal cells do not. (**1**)

3   Panther / *Panthera pardus* and wolf / *Canis lupus* (**1**); because they both belong to the same (kingdom, phylum, class and) order (**1**)

4

Domain	Distinguishing characteristic of the domain
Archaea	**cells with no nucleus, genes contain unused sections of DNA**
Eubacteria	**cells with no nucleus, no unused sections in genes**
Eukarya	cells with a nucleus, unused sections in genes

(**1 mark** for each correct row)

42.   Selective breeding

1   (a) Plants or animals with certain desirable characteristics are chosen to breed together (**1**) so that their offspring will inherit these characteristics (**1**).

(b) pigs with lower body fat are crossed (**1**); offspring with low body fat are selected and crossed (**1**); repeated for many generations until a lean breed is produced (**1**)

2   (a) high yield so can feed more people (**1**); low fertiliser requirement so no need to apply fertiliser / reduce cost (**1**); pest resistant (or example given) so less pest damage / do not need to apply pesticide (**1**)

(b) drought resistant to cope with times of water shortage without dying (**1**); tolerant of high temperature (**1**)

(c) less likely to be blown over in the wind / less likely to snap / plant uses less energy in growing the stem, so has more to use for making seeds (**1**)

3   Three from: alleles that might be useful in the future might no longer be available (**1**); a new disease might affect all organisms (**1**); selectively bred organisms might not adapt to changes in climate (**1**); animal welfare might be harmed (**1**)

43.   Genetic engineering

1   Mice do not normally glow, but glow mice have a gene from jellyfish inserted into their DNA / genome (**1**) that produces / codes for a protein that glows in blue light (**1**).

2   (a) Rice / tomatoes / wheat (**1**)

(b) Two from: increased yields (1) increased nutritional value (1) resistance to attack by insects (1) resistance to herbicides (so that weeds are killed but not the crop) (1), resistance to drought (1)

(c) Disadvantage (1) with reason (1), e.g. may kill insect species other than pests, so loss of diversity; insects are food for birds, so less food for birds; gene may transfer to another plant (such as a weed) so more of these plants grow amongst the crop plants.

3 Advantages: can manufacture very pure product (1); can manufacture large quantities of insulin (1); as insulin is human insulin, few problems with rejection (1); overcomes need to harvest insulin from animals (1) Disadvantages: ethical issues over modification of organisms using human genes (1); possibility that GM bacteria may prove unsafe / harmful in the long term (1)

(1 mark for each valid point. Full marks are only available if at least one advantage and one disadvantage is given.)

44. Tissue culture

1 (a) Take a small piece of tissue from the parent plant (1); place this in agar growth medium containing nutrients and auxins (1); the tissue develops into tiny plantlets (1) which are planted into compost (1).

(b) Many genetically identical plants can be produced quickly (1), all with the desirable characteristics of the parent plant (1).

(c) One from: if there is an undesirable characteristic, all the plants will have it (1); plant hormones are needed (1); growth medium / nutrients are needed (1); process requires skill (1)

2 (a) Obtain human (stem) cells (1) and grow in liquid containing nutrients (1).

(b) Cell culture will be quicker (1) and cheaper (1) and does not have any ethical (1) issues if the drug damages the cells. Also, human cells could be used which may respond more like humans / patients (1).

45. Stages in genetic engineering

1 (a) a small circle of DNA from bacteria (1)

(b) something that carries a new gene into a cell (1)

(c) a few unpaired bases at the ends of double-stranded DNA (1)

2 (a) The human gene needed is the one for insulin (1). It is needed because it codes for the protein / hormone (1).

(b) Enzymes have two roles – these are to cut the human gene out of the chromosome (1) and insert it into the bacterial DNA (1).

(c) The bacteria provide plasmid (1) DNA for the process; the bacteria are useful because they produce large quantities of human insulin (1).

3 (a) The pieces of human DNA have sticky ends (1) and the plasmids have matching sets of unpaired bases (1) so that the bases in the pieces of human DNA pair up with the bases in the plasmids (1).

(b) DNA ligase links the human DNA and plasmid back into a continuous circle (1) so that the plasmids can be inserted back into the bacteria (1).

46. Insect-resistant plants

1 (a) The Bt toxin is a substance that is poisonous / kills (1) insects / named insect pests, e.g. caterpillars (1), so it reduces damage to plants (1).

(b) (One type of bacterium) contains / supplies the Bt gene (1); (a different bacterium) has the Bt gene added to its plasmid (1) and this acts as a vector / adds the gene to plant cells (1).

2 (a) increased yield (1) because of less damage by insects / pests (1) OR less chemical insecticide is needed (1) so other harmless / useful insects are less likely to be harmed (1)

(b) (i) Pollen transferred from insect-resistant / transgenic plant (1) could fertilise the eggs inside the wild plant flowers causing transfer of gene to offspring (1).

(ii) Insects that feed on the wild plants could be killed (1); because they would eat the Bt toxin (1).

47. Meeting population needs

1 (a) use of artificial fertilisers (1)

(b) use of biological control (1)

2 (a) The population of aphids increases slowly at first (1) and then very rapidly (1), reaching 30 aphids per leaf after 10 days (1).

(b) When ladybirds are present, the number of aphids increases more slowly / does not rise so quickly / is less than without the ladybirds (1) because ladybirds eat aphids (1) but cannot eat them all / some aphids still survive to breed (1).

3 Two from: biological control takes time to act as control agents reproduce (1); control agents might become pests (1); less pesticide is needed when used with biological control (1); chemical pesticides get rid of the pest completely (1).

4 EITHER: The advantages outweigh the disadvantages because fertilisers increase crop yields (1) so allow us to grow more food (1); however, they are expensive / can cause pollution / reduce soil biodiversity (1). OR: The advantages do not outweigh the disadvantages because artificial fertilisers are expensive (1) or can cause pollution (1) or reduce soil biodiversity (1) even though they increase crop yields (1).

(Marks awarded for giving two valid points in favour of the conclusion and one against.)

48. Extended response – Genetic modification

*Answer could include the following points:

- Both rely on variation caused by mutation and sexual reproduction / meiosis.
- Evolution involves natural selection.
- Variation means some individuals are better able to survive in their environment.
- They will produce more healthy offspring than others.
- So their alleles are more likely to be passed on.
- Also explains how organisms adapt to changes in their environment over several generations.
- Selective breeding involves artificial selection.
- Plants or animals with desired characteristics are bred together.

- Offspring will inherit these characteristics.
- Those with the most desirable characteristics are bred further.
- Repeated many times.
- Does not involve adaptation to changed environment.
- But can be used to breed plants / animals better able to cope with difficult conditions.

49. Health and disease

1 (a) being free from disease and eating and sleeping well (1)

(b) how you feel about yourself (1)

(c) how well you get on with other people (1)

2 (a) Communicable: ✓influenza, ✓tuberculosis; ✓Chlamydia; Non-communicable: ✓lung cancer, ✓coronary heart disease

(3 marks for 5 correct, 2 marks for 3 or 4 correct, 1 mark for 1 or 2 correct)

(b) Communicable: rapid variation in number of cases over time / cases localised (1); non-communicable: number of cases change gradually / cases more wide spread (1)

3 (HIV) causes damage to the immune system (1) reduced immune response / immunity (1)

4 (a) Three from: a virus infects a body cell (1) and takes over the body cell's DNA (1) causing the cell to make toxins (1) or damages the cell when new viruses are released (1).

(b) Bacteria can release toxins (1) and can invade and destroy body cells (1).

50. Common infections

1 (a) Zimbabwe (1); 15.1–14.3 = 0.8% decrease (1)

(b) All countries show a decrease in the % of 15 to 49 year olds with HIV (1), one example of a trend such as: all % have dropped somewhere between 0.3 and 2.9% (1).

2 (a) D (1)

(b) Two from: leaf loss; (1); bark damage (1); dieback of top of tree (1)

3

Disease	Type of pathogen	Signs of infection
cholera	**bacterium**	watery faeces
tuberculosis	bacterium	persistent cough – may cough up blood
malaria	**protist**	**fever, weakness, chills and sweating**
HIV	**virus**	mild flu-like symptoms at first

(all correct for 3 marks, 3 correct for 2 marks, 2 or 1 correct for 1 mark)

4 (a) Bacterium (1)

(b) Two from: inflammation in stomach (1) bleeding in stomach (1) stomach pain (1)

51. How pathogens spread

1 C (**1**)

2

Disease	Pathogen	Ways to reduce or prevent its spread
Ebola haemorrhagic fever	**virus (1)**	keep infected people isolated; wear full protective clothing while working with infected people or dead bodies
tuberculosis	bacteria	**ventilate buildings to reduce chance of breathing in bacteria / diagnose promptly and give antibiotics to kill bacteria / isolate infected people (1)**

3 Boil water before drinking / wash hands after using toilet (**1**) because bacteria are spread in water / by touch (**1**).

4 (a) The bacteria are spread in water (**1**); in developed countries water is treated to kill pathogens / good hygiene prevents their spread (**1**).

   (b) To prevent being infected by the Ebola virus (**1**) because Ebola virus is present in body fluids of infected people even after death (**1**).

52. STIs

1 An infection spread by sexual activity (**1**)

2 B (**1**)

3

Mechanism of transmission	Precautions to reduce or prevent STI
unprotected sex with an infected partner	using condoms during sexual intercourse (**1**)
sharing needles with an infected person (1)	supplying intravenous drug abusers with sterile needles
infection from blood products	**screening blood transfusions (1)**

4 (a) Viral genetic material inserts itself into the cell's genetic material (**1**) and is replicated with the cell's genetic material (**1**).

   (b) because HIV can spend many years in the lysogenic cycle (**1**) and AIDS only develops when HIV enters the lytic cycle (**1**)

53. Human defences

1 (a) Skin acts a physical barrier that stops microorganisms getting into the body (**1**).

   (b) Hydrochloric acid in the stomach kills pathogens (**1**).

   (c) (i) lysozyme (**1**)

       (ii) kills bacteria (**1**) by digesting their cell walls (**1**)

2 (a) (i) Mucus (**1**)

       (ii) Sticky so traps bacteria / pathogens (**1**)

   (b) (i) Cilia (**1**)

       (ii) The cilia on the surface of these cells move in a wave-like action (**1**) and this moves mucus and trapped pathogens out of lungs (**1**) towards the back of the throat where it is swallowed (**1**).

   (c) Mucus travels down to into the lungs carrying pathogens (**1**) because the cilia cannot move and take the pathogens back up to the throat (**1**)

54. The immune system

1 lymphocytes (**1**)

2 Pathogens have substances called antigens (**1**) on their surface. White blood cells called lymphocytes (**1**) are activated if they have antibodies (**1**) that fit these substances. These cells then divide many times to produce clones / identical cells (**1**). They produce large amounts of antibodies that stick to the antigens / destroy the pathogen (**1**).

3 (a) Lymphocytes producing antibodies against measles virus are activated (**1**); these lymphocytes divide many times (**1**); so concentration of antibodies increases (**1**) then decreases when the viruses have all been destroyed (**1**).

   (b) Some of the lymphocytes stay in the blood as memory lymphocytes (**1**); these respond / divide after infection (**1**) so the number of lymphocytes producing the antibodies against the measles virus increases rapidly (**1**).

   (c) (i) (The girl had not been exposed to the chicken pox virus before because) line B is similar in size and shape to line A (**1**) which was for a first infection with measles / the line would be higher if it was a second infection (**1**).

       (ii) The concentration of antibodies increased faster / to a higher concentration (**1**) so the measles viruses were destroyed before they could cause illness / symptom / disease (**1**).

55. Immunisation

1 (a) A vaccine contains antigens from a pathogen (**1**), that are inactive / unable to cause disease (**1**).

   (b) The vaccine causes memory lymphocytes to be produced (**1**), so if the person is exposed to the disease the memory lymphocytes produce a secondary response (**1**); this prevents an increase in the pathogen to a level that causes illness (**1**).

   (c) Two suitable such as: mild swelling (**1**); soreness (**1**); a mild form of the disease (**1**); rare major harmful reaction (**1**)

2 (a) 2003 (**1**)

   (b) The number of cases would increase (**1**) because fewer babies were immunised which would protect them from infection (**1**).

   (c) More babies would become immune (**1**) so unvaccinated babies would be less likely to catch measles / herd immunity (**1**).

56. Treating infections

1 (a) C (**1**)

   (b) Antibiotics kill bacteria / inhibit their cell processes (**1**) but do not affect human cells (**1**).

2 The pharmacist's advice would be not to take the penicillin (**1**). The man's cold is due to a virus, so the penicillin will not be effective in combating the infection (**1**).

3 (a) Sinusitis is (probably) not caused by a bacterial infection (**1**).

   (b) Same number of patients got better without antibiotic (**1**) although the patients taking antibiotic may have got better more quickly (**1**).

57. Aseptic techniques

1 (a) so that microorganisms on the workbench are killed / do not contaminate the culture (**1**)

   (b) One from: because the microorganisms need oxygen to grow (**1**); because lack of oxygen encourages growth of pathogenic microorganisms (**1**)

   (c) because pathogenic microorganisms are more likely to grow at 37 °C / body temperature (**1**)

   (d) so there is no risk of infection / cross-contamination (**1**)

2 (a) This will stop the entry of microorganisms from the air (**1**) that are unwanted / not part of the experiment / likely to contaminate the Petri dish (**1**).

   (b) This sterilises the loop (**1**) to prevent unwanted microorganisms getting cultured (**1**).

   (c) This stops other microorganisms getting in (**1**) and contaminating the culture (**1**).

3 Three from: heating the jelly to 80 °C kills any bacteria or other microorganisms in the jelly (**1**); cooled to 21 °C reduces the risk of harmful bacteria being present in the culture (**1**); warming to a higher temperature produces a more rapid growth (**1**); using sterilised Petri dishes prevents potentially harmful bacteria contaminating the bacterial culture (**1**)

58. Investigating microbial cultures

1 (a) bacteria do not grow in clear area (**1**) because the antibiotic kills them / stops them growing (**1**)

   (b)

Antibiotic	Diameter of clear area (mm)	Cross-sectional area (mm²)
1	7	38.5
2	11	95.0
3	12	113.1
4	10	78.5

   1 mark for each diameter and area

   (c) Disc 3 (**1**); killed the largest area of bacteria (**1**).

2 (a) Use different concentrations of antibiotic (**1**) but the same volume of each (**1**). Repeat with several identical plates and

calculate the mean diameter for each concentration (**1**).

(b) Two from: discs are the same size (**1**); same volume of antibiotic solution used (**1**); discs left on the dish for the same length of time (**1**)

59. New medicines

1 (a) 3, 1, 5, 2, 4 (all correct = **2 marks**, 4 correct = **1 mark**)

(b) (i) Testing in cells or tissues to see if the medicine can enter cells and have the desired effect (**1**), testing on animals to see how it works in a whole body / has no harmful side effects (**1**).

(ii) by testing in a small number of healthy people (**1**)

(c) Medicine is tested on people with the disease that it will be used to treat (**1**) so that the correct dose can be worked out (**1**) and to check for side effects in different people (**1**).

2 (a) Large number of subjects make the data valid (**1**) and repeatable (**1**); OR side effects will only be seen in small numbers (**1**) so it is easier to notice with a large trial group (**1**); OR there are different stages of the trial (**1**) and each step needs a different group of people (**1**).

(b) The medicine appears to be effective in nearly 400 people with high blood pressure (**1**); this reduction is much greater than those in the placebo group (**1**). You could also say: the medicine seems to have very little adverse effect on the blood pressure of those in the 'normal' group (so it is effective).

60. Monoclonal antibodies

1 (a) lymphocyte (**1**)

(b) Once they start producing antibodies they stop dividing (**1**) so it is difficult to produce many antibody-producing cells (**1**).

2 (a) antibodies of one type (**1**) produced in large quantities by hybridoma cells (**1**).

(b) Any four from: mouse is injected with the pregnancy hormone (**1**); the hormone acts as an antigen (**1**); the mouse produces lymphocytes that make antibodies to the hormone (**1**); the lymphocytes are fused with cancer cells (**1**); to form hybridoma cells and cultured to make large amounts of monoclonal antibodies to the hormone (**1**).

(c) The antibody / hormone is not the correct shape to recognise a different hormone / antibody (**1**).

3 (a) Two from: Monoclonal antibodies are made that attach to antigens only present on cancer cells (**1**). The antibodies are made slightly radioactive (**1**). The position of the antibodies is detected using a PET scanner (**1**).

(b) The cancer medicine is attached to a monoclonal antibody (**1**) that only binds to / recognises antigens on cancer cells (**1**), so the medicine is delivered to cancer cells (**1**) and less is delivered to healthy cells that don't have the antigen (**1**).

61. Non-communicable diseases

1 An infectious disease is caused by a pathogen (**1**) and is passed from one person to another (**1**). A non-communicable disease is not passed from one person to another (**1**).

2 Three from: inherited / genetic factors (**1**); age (**1**); sex (**1**); ethnic group (**1**); lifestyle (e.g. diet, exercise, alcohol, smoking) (**1**); environmental factors (**1**)

3 (a) (i) Bangladeshi men (**1**)

(ii) black women (**1**)

(b) Four from: the prevalence of CHD increases with age (**1**); overall the prevalence is higher in men than in women (**1**), but prevalence is similar in black men and women (**1**); Bangladeshi men have the highest prevalence but Bangladeshi women are in the middle (**1**); ethnic group seems to be a bigger factor in men than in women (**1**); the prevalence in all ethnic groups is very similar in the 40–49 age group (**1**)

62. Alcohol and smoking

1 (a) Ethanol is a drug that is toxic / poisonous to cells (**1**). It is broken down by the liver and harms liver cells (**1**). Too much alcohol over a long period causes cirrhosis / liver disease (**1**).

(b) because it is caused by how we choose to live (**1**)

2 Two from: because carbon monoxide in cigarette smoke (**1**) reduces how much oxygen the blood can carry to the baby (**1**), leading to low birth weight in babies / other abnormalities (**1**).

3 (a) Two from: cardiovascular disease (**1**); lung cancer (**1**); respiratory / lung disease (**1**)

(b) Substances in cigarettes cause blood vessels to narrow (**1**) which increases the blood pressure (**1**) leading to cardiovascular disease (**1**).

63. Malnutrition and obesity

1 (a) too little of one or some nutrients in the diet (**1**)

(b) Four from: anaemia increases with increasing age (**1**) in both men and women (**1**), but whereas there is an increase in females from 1–16 and 17–49 (**1**) followed by a decline (**1**), in males the lowest age groups are 17–49 and 50–64 (**1**).

2 (a)

Subject	Weight (kg)	Height (m)	BMI
Person A	80	1.80	24.7
Person B	90	1.65	33.1
Person C	95	2.00	23.8

All 3 correct = **2 marks**, 2 correct = **1 mark**

(b) person B (**1**)

3 Too much fat / obesity increases the risk of cardiovascular disease (**1**) and abdominal fat is most closely linked with cardiovascular disease (**1**); measuring waist : hip ratio is a better measure of abdominal fat (**1**).

64. Cardiovascular disease

1 (a) Two from: lifestyle changes (**1**); medication (**1**); surgery (**1**)

(b) Two from: give up smoking (**1**); take more exercise (**1**); eat a healthier diet (lower fat, sugar and salt) (**1**); lose weight (**1**)

(c) because cardiovascular disease reduces life expectancy (**1**) and can be fatal before treatment can be given (**1**)

2 Lifestyle changes – Benefits: no side effects; may reduce chances of getting other health conditions / the cheapest option. Drawbacks: may take time to work; may not work effectively.

Medication – Benefits: easier to do than change lifestyle; starts working immediately / cheaper and less risky than surgery. Drawbacks: can have side-effects; needs to be taken long-term / may not work well with other medication the person is taking.

Surgery – Benefits: once recovered, there are no side effects / usually a long term solution. Drawbacks: risk of infection after surgery; there is a risk the person will not recover after the operation / expensive / more difficult to do than giving medication.

(**3 marks** for 6 correct, **2 marks** for 4 or 5 correct, **1 mark** for 2 or 3 correct)

3 Surgery can help prevent heart attacks / strokes (**1**) but costs more than inserting a stent (**1**) and surgery has more risk (e.g. risk of infection). (**1**) However, it can be a long-term solution / other suitable conclusion (**1**).

65. Plant defences

1 (a) Two from: bark (**1**); thick waxy cuticles (**1**); spikes / thorns (**1**); cellulose cell walls (**1**)

(b) One from: produce poisons to kill pests / pathogens (**1**); produce chemicals to stop infection (**1**)

(c) the control (**1**)

(d) To stop other bacteria from getting in (**1**); to stop organisms entering that would kill the bacteria (**1**).

(e) Two from: temperature (**1**); same volume of garlic juice and water (**1**); same species of bacteria (**1**); same number of bacteria (**1**); same amount of jelly in each tube (**1**)

(f) The plant is able to defend itself from attack by pathogens (**1**).

2 (a) Two from: toxic chemicals (**1**); bitter tasting chemicals (**1**); spines on leaves to deter feeding (**1**)

(b) to kill pathogens (**1**) to treat symptoms / inflammation (**1**)

66. Plant diseases

1 (a) One from: drought (**1**), waterlogging (**1**), cold / heat (**1**), nutrient supply (**1**)

(b) Two from: change in normal appearance (**1**), overgrowth of part of the plant (**1**), under-development of part of the plant (**1**), death of part of the plant (**1**)

(c) different diseases may cause the same symptoms (**1**)

Answers

(d) (i) so that the soil could be sent for analysis **(1)** to look for soil factors / nutrient deficiency / pH / toxins **(1)**

(ii) to test for the presence of pathogens **(1)** by microscopy / genetic analysis / growing pathogen on nutrient agar **(1)**

2 *Answer to include four of the following points:* Take samples from affected trees and examine with a microscope for signs of pathogens **(1)**; use antibodies to test for the presence of a pathogen **(1)**; use genetic testing to identify any pathogens found **(1)**; use soil sample testing to rule out soil factors, e.g. nutrient deficiency **(1)**; try to grow pathogen on nutrient medium to produce a larger sample for identification **(1)**.

67. Extended response – Health and disease

* Answer could include the following points:

- Antibiotics kill bacteria or inhibit their cell processes and stop them growing.
- Antibiotics can be given after a person is infected with a pathogen.
- Antibiotics are not effective against viruses.
- Bacteria can become resistant to antibiotics.
- Antibiotics can have side-effects.
- Vaccines trigger the body's own immune system.
- Vaccines must be given before a person becomes infected.
- Vaccines can protect against viruses as well as bacteria.
- Herd immunity means not everyone needs to be vaccinated.
- In rare cases a person may react badly to the vaccine.

68. Photosynthesis

1 Plants or algae are photosynthetic organisms / producers **(1)** so they are the main producers of biomass **(1)** and animals have to eat plants / algae **(1)**.

2 (a) carbon dioxide + water **(1)** → glucose + oxygen **(1)**

(b) The product of photosynthesis / glucose has more energy than the reactants **(1)** because energy is transferred from the surroundings / light **(1)**.

3 (a) Light is required for photosynthesis **(1)** because only parts of the leaf exposed to light produced starch **(1)**.

(b) Chlorophyll / chloroplasts required for photosynthesis **(1)** because only green areas / areas with chlorophyll or chloroplasts produce starch **(1)**.

69. Limiting factors

1 (a) temperature **(1)**
(b) Add algal balls to hydrogen carbonate solution **(1)**. Leave for a set amount of time e.g. 2 hours **(1)**. Compare the colour change against standard colours **(1)**.

2 (a) Increasing the carbon dioxide concentration increases the rate of photosynthesis **(1)**.

(b) Adding carbon dioxide means it will no longer be a limiting factor **(1)** so the plants can make more sugars needed for growth **(1)**.

(c) You could the increase temperature **(1)** as this would make photosynthesis happen faster / more quickly **(1)**.

3 Increased temperature increases the rate of reaction **(1)** so photosynthesis / growth happens faster **(1)**; eventually other factors limit rate / rate reaches maximum **(1)**; higher temperatures denature enzymes responsible for photosynthesis **(1)**.

70. Light intensity

1 (a) points plotted accurately **(1)** curve of best fit drawn **(1)**

(b) 76 (±2) **(1)**

(c) the greater the light intensity, the higher the rate **(1)**; not a linear relationship **(1)**

(d) (i) Take care not to touch the bulb if it is hot **(1)**.

(ii) Place a water tank next to the bulb if it is hot **(1)** to help prevent heat from the lamp reaching the test tube **(1)**. OR Use a ruler to make sure that the lamp is at the measured distance **(1)** because differences in distance will change the light intensity **(1)**.

(e) You could use the light meter to measure the light intensity **(1)** at each distance and then plot a graph of rate of photosynthesis bubbling against light intensity **(1)**.

71. Specialised plant cells

1 (a) phloem **(1)**

(b) A There are holes **(1)** to let liquids flow from one cell to the next **(1)**.

B There is small amount of cytoplasm so **(1)** there is more space for the central channel **(1)**.

(c) Mitochondria supply energy **(1)** for active transport (of sucrose) **(1)**.

2 (a) xylem **(1)**

(b) Three from: the walls are strengthened with lignin rings to prevent them from collapsing **(1)**; no cytoplasm means there is more space for water **(1)**; pits in the walls allow water and mineral ions to move out **(1)**; no end walls means they form a long tube so water flows easily **(1)**

72. Transpiration

1 (a) Transpiration is the loss of water **(1)** by evaporation from the leaf surface **(1)**.

(b) stomata (in the leaf) **(1)**

(c) (i) moves faster **(1)** because a faster rate of water loss from leaves **(1)**

(ii) moves slower **(1)**; stomata covered so a lower rate of water loss **(1)**

2 (a) Guard cells take in water by osmosis **(1)** so they swell, causing the stoma to open **(1)**; when the guard cells lose water they become flaccid / lose rigidity and the stoma closes **(1)**.

(b) The stomata are open during the day, so water is lost by transpiration **(1)** faster than it can be absorbed by the roots **(1)**. Water is lost from the vacuoles and the plant wilts. At night, the stomata close so water is replaced **(1)**.

73. Translocation

1 (a) the movement of sucrose around a plant **(1)**

(b) A **(1)**

2 (a) Radioactive carbon dioxide is supplied to the leaf **(1)**. Radioactive carbon / sucrose will then be detected in the phloem **(1)** and eventually incorporated into starch in the potato **(1)**.

(b) Radioactivity would remain in the leaf / not appear in the phloem **(1)** because companion cells actively pump sucrose into / out of the phloem **(1)**.

3

Structure or mechanism	Transport of water	Transport of sucrose
Xylem	X	
Phloem		X
Pulled by evaporation from the leaf	X	
Requires energy		X
Transported up and down the plant		X

(**1 mark** for each correct row.)

74. Leaf adaptations

1 (a) (i) guard cell(s) **(1)**

(ii) Guard cells open stomata during the day and close at night **(1)**; this allows CO_2 to enter during the day and reduces water loss at night **(1)**.

(b) (i) It is thin / transparent **(1)** to allow more light to pass through **(1)**.

(ii) contain a lot of chloroplasts / packed closely together / cylindrical shape **(1)** to maximise locations where photosynthesis can occur **(1)**

(c) Internal air space increases surface area **(1)** to increase rate of diffusion of gases **(1)**.

2 (a) Large leaves also have a large surface area **(1)** so they can absorb more light for photosynthesis **(1)**.

(b) Xylem vessels bring water to the leaf **(1)**; phloem transports sugar away from the leaf **(1)**.

75. Water uptake in plants

1 (a) Rate of transpiration increases **(1)** because increasing light intensity causes stomata to open **(1)**.

(b) Rate of transpiration increases **(1)** because higher temperature increases energy of water molecules / move faster **(1)**.

2 (a) The rate of evaporation was higher when the fan was on **(1)**; because the movement of air removes water more quickly from the leaves **(1)**, increasing the concentration gradient from leaf to air **(1)**.

(b) The rate of evaporation became quicker than the rate at which the plant could take up water **(1)**; the stomata of the plant closed **(1)** to prevent evaporation from occurring / conserve water **(1)**.

(c) The volume of the tube is calculated using $\pi r^2 l$. Volume of 90 mm length of tube = $(\pi \times 0.25^2 \times 90) \div 5$ **(1)** = 3.53. Rounded to 1 dp = 3.5 mm³ / min **(1)**

76. Plant adaptations

1 (a) The lianas' leaves are high up where there is light / more light **(1)** but the roots are in the ground where it is very wet **(1)**.

(b) EITHER large leaves **(1)** to take in as much light as possible **(1)** OR drip tips on the leaves **(1)** to allow water to run off **(1)**

2 (a) Rate of evaporation of water from leaves is reduced **(1)** because leaves have small surface area to volume ratio **(1)**.

(b) (i) waxy cuticle is waterproof **(1)**; less water lost through upper surface of leaf **(1)**

(ii) deep pits and hairs trap water vapour **(1)**; so rate of diffusion of water out of leaf is lower **(1)**

3 Three from: large leaves absorb more light for photosynthesis **(1)**; they lose water rapidly but there is rainfall / more water in the spring **(1)**; small leaves have lower rates of evaporation **(1)** so conserve water in summer **(1)**.

77. Plant hormones

1 (a) X on tip of shoot **(1)**

(b) Y on area just behind the tip on the left hand side **(1)**

(c) Z on the vertical part of the root **(1)**

2 (a) Phototropism causes shoots to bend towards the light **(1)**. This means leaves are positioned better to capture more sunlight for photosynthesis **(1)**.

(b) It causes plant roots to grow downwards **(1)** where there is more water / helps to anchor the plant into the ground **(1)**.

3 Auxins move from the tip of the shoot **(1)** to the shaded side of the plant **(1)**; cells on the shaded side elongate **(1)**.

78. Uses of plant hormones

1 (a) Two from: stimulate germination **(1)**; stimulate flower / fruit production **(1)**; increase stem length in sugar cane **(1)**

(b) Gibberellins stop seeds developing **(1)** and fruit grow larger **(1)**; people pay more for large seedless fruit **(1)**.

(c) Spray plants with gibberellin **(1)** that overrides photoperiodism so plants flower early **(1)**.

2 Food does not become overripe when travelling long distances from other countries **(1)** and has a longer shelf life / reaches shops in 'just-ripened' condition **(1)**.

3 Artificial auxins kill broad leaf plants **(1)** so do not affect narrow leaf crops **(1)**.

4 Auxins in rooting powders cause plant cuttings to develop roots quickly **(1)**; large numbers of identical plants can be produced quickly **(1)**.

79. Extended response – Plant structures and functions

*Answer could include the following points:

• Stomata allow carbon dioxide from air to enter leaf and oxygen to leave.

• Internal air spaces increase area for diffusion of gases.

• Xylem cells bring water needed for photosynthesis.

• All of these adaptations can increase water loss.

• Water loss is greatest at high temperature / light intensity and dry / windy conditions.

• Plants growing in dry conditions are adapted to reduce water loss.

• Waxy cuticle and stomata sunk in pits reduce water loss.

• Rolled leaves reduces air movements around stomata.

• Leaf hairs trap moist air around stomata.

80. Hormones

1 (a) Hormones are produced by endocrine glands **(1)** and are released into the blood **(1)**. They travel round the body until they reach their target organ **(1)**, which responds by releasing another chemical substance **(1)**.

(b) Hormones have long-lived effects; nerves have short-term effects **(1)**. Nerve impulses act quickly; hormones take longer **(1)**.

2 A = hypothalamus **(1)**, B = pituitary **(1)**, C = thyroid **(1)**, D = pancreas **(1)**, E = adrenal **(1)**, F = testis **(1)** and G = ovary **(1)**.

3

Hormone	Produced in	Site of action
TRH	**hypothalamus**	pituitary gland
TSH	pituitary gland	**thyroid gland**
ADH	pituitary gland	**kidney**
FSH and LH	**pituitary gland**	ovaries
insulin and glucagon	**pancreas**	liver, muscle and adipose fatty tissue
adrenalin	**adrenal gland**	various organs, e.g. heart, liver, skin
progesterone	**ovaries**	uterus
testosterone	**testes**	male reproductive organs

All 8 correct = **4 marks**, 6 or 7 correct = **3 marks**, 4 or 5 correct = **2 marks**, 2 or 3 correct = **1 mark**.

81. Adrenalin and thyroxine

1 (a) Two from: heart beats faster **(1)** so oxygen is carried around the body faster **(1)** OR some blood vessels constrict **(1)** so blood pressure increases **(1)** OR some blood vessels dilate **(1)** so blood flow to muscles increases / delivery of oxygen and nutrients to muscles increases **(1)** OR liver converts glycogen to glucose **(1)** so more glucose available for respiration **(1)**

(b) An increase in thyroxine concentration causes changes **(1)** that bring about a decrease in the amount of thyroxine released **(1)** (or vice versa).

(c) When blood concentration of thyroxine is lower than normal **(1)** hypothalamus produces TRH **(1)** which stimulates pituitary to produce TSH **(1)** which

stimulates thyroid gland to produce thyroxine **(1)** (or reverse if thyroxine concentration is higher than normal).

2 Thyroxine controls the resting metabolic rate **(1)** so is produced constantly **(1)**. Adrenalin is produced in response to fright / excitement **(1)** and is only produced when needed **(1)**.

82. The menstrual cycle

1 Two from: oestrogen **(1)**; progesterone **(1)**; FSH **(1)**; LH **(1)**

2 (a) A = menstruation **(1)**; B = ovulation **(1)**

(b) any time between day 14 and about day 17 **(1)**

(c) The lining of the uterus breaks down **(1)** and is lost in a bleed or period **(1)**.

3 (a) Pills, implants or injections release hormones that prevent ovulation **(1)**, and thicken mucus at the cervix **(1)**, preventing sperm **(1)** from passing.

(b) (i) These figures are maximum values **(1)** when the methods are used correctly **(1)**.

(ii) hormonal method most effective in preventing pregnancy **(1)**; diaphragm / cap least effective in preventing pregnancy **(1)**; only condom offers protection against STIs **(1)**

83. Control of the menstrual cycle

1 (a) pituitary gland **(1)**, ovary / mature follicle **(1)**

(b) corpus luteum **(1)**, pituitary gland / lining of uterus **(1)**

(c) ovary / maturing follicle **(1)**, pituitary gland **(1)**

2 (a) (i) Levels of progesterone are low **(1)** which allows FSH release by the pituitary gland **(1)**.

(ii) Level of LH rises rapidly / LH surge **(1)** triggers ovulation **(1)**.

(iii) Progesterone levels fall after day 23 **(1)** so uterus wall thickness is not maintained and therefore pregnancy has not occurred **(1)**.

(b) High level of progesterone **(1)** prevents release of FSH **(1)** so follicles aren't stimulated to grow **(1)**.

(c) Increase in temperature coincides with ovulation **(1)** so she will be more fertile in the days after **(1)**.

84. Assisted Reproductive Therapy

1 (a) Clomifene helps increase concentration of FSH and LH in blood **(1)** so stimulates maturation of follicles and release of eggs **(1)**.

(b) Clomifene will stimulate release of eggs **(1)** but these cannot reach the uterus if oviducts are blocked **(1)**.

2 (a) FSH stimulates maturation of follicles **(1)**; FSH stimulates release of many eggs **(1)**.

(b) eggs and sperm mixed to allow fertilisation **(1)**; one or two healthy embryos placed in uterus **(1)**

(c) cell can be removed from embryo before being placed in the uterus **(1)**; this can be tested for genetic disorders **(1)**

3 Four from the following, but there must be at least one drawback and at least one advantage.

Drawbacks: the success rate for IVF is still quite low (**1**); at only 12 400 / 45 250 × 100 = 27% (**1**); cost is quite high (especially if more than one cycle of treatment is needed) (**1**); total cost to the NHS in 2010 = 45 250 × £2500 = £113 million (**1**)

Benefits: it allows couples to have children if they are not able to naturally (**1**); even though success rate is low, it is still successful for some couples (**1**); it can be easier to use this procedure than to adopt (**1**)

85. Homeostasis

1 maintaining conditions inside the body at a more or less constant level (**1**) in response to internal and external changes (**1**)

2 (a) the hypothalamus (**1**)

(b) Enzymes have an optimum temperature (**1**) so the body temperature must be kept at this level (**1**).

(c) Shivering means energy is released from respiration (**1**) which increases the core body temperature (**1**).

3 (a) to control the amount of water in the body (**1**) by controlling how much water is lost in the urine (**1**)

(b) to stop the body cells from swelling up (**1**) by absorbing water by osmosis (**1**)

4 Because the amount of water in the body has to rise above / fall below the normal level (**1**) before the brain detects the change (**1**) and takes action to correct it (**1**).

86. Controlling body temperature

1 (a) muscle contracts when cold to pull hairs upright (**1**) trapping an insulating layer of air (**1**); when warm the muscle relaxes and hairs lie flat (**1**)

(b) sweat glands secrete sweat onto the skin / epidermis (**1**); this evaporates cooling the body down / taking heat away from the body (**1**)

2 (a) core body temperature remains stable / does not change (**1**); temperature at skin's surface rises slightly / rises to a maximum of 38°C (**1**)

(b) When the temperature rises blood vessels dilate (**1**). There is greater flow of blood to the skin's surface (**1**). This means that more heat radiates from the surface of the skin (**1**).

(c) When they are cold less blood flows near the surface of the skin (**1**) because of vasoconstriction (**1**).

87. Blood glucose regulation

1 order is: 3, 5, 1, 4, 2. All 5 correct = **3 marks**, 3 correct = **2 marks**, 1 correct = **1 mark**.

2 (a) A&F pancreas (**1**), B insulin (**1**), C&D liver (**1**), E glucagon (**1**)

(b) (i) Insulin causes the liver to take up excess glucose and convert it to glycogen (**1**), causing blood glucose concentration to fall (**1**).

(ii) Glucagon causes the liver to convert glycogen to glucose, which is released into the blood (**1**), causing the blood glucose concentration to rise (**1**).

88. Diabetes

1 (a) As the BMI increases the percentage of people with diabetes increases (**1**) so there is a positive correlation (**1**).

(b) (i) BMI = 88 ÷ 1.8² = 27.2 (**1**); he (is overweight so) has an increased risk of Type 2 diabetes (**1**) but not the highest risk (**1**).

(ii) waist : hip ratio = 104 ÷ 102 = 1.02 which is obese (**1**) so he has a high risk of developing Type 2 diabetes (**1**) because there is a correlation between waist : hip ratio and risk of Type 2 diabetes (**1**).

2 (a) Controlling diets will help to control the number of people who are obese (**1**). Fewer obese people means fewer with diabetes (**1**).

(b) (i) In Type 1 diabetes no insulin is produced so has to be replaced with injections (**1**) but in Type 2 diabetes organs don't respond to insulin (**1**).

(ii) because a large meal means a higher blood glucose concentration (**1**) so more insulin is needed to reduce the glucose concentration (**1**)

89. The urinary system

1 (a) A kidney; B renal artery; C renal vein; D ureter; E bladder; F urethra; (**1 mark** for every 2 correct)

(b) (i) urea (**1**)

(ii) liver cells (**1**)

(c) A: removes excess amounts of some substances / urea and makes urine (**1**); B: carries blood from body to kidneys (**1**); D: carries urine from kidneys to the bladder (**1**); muscle: keeps exit from the bladder closed until urination (**1**)

2 (a) (i) Proteins stay in the blood (**1**) because they are too large to pass into the nephron (**1**).

(ii) Glucose is small so it passes into the nephron (**1**) but is reabsorbed by active transport in the first convoluted tubule (**1**).

(iii) Urea is small so it passes into the nephron (**1**) but is not reabsorbed (**1**); its concentration in urine is higher because water is reabsorbed (**1**).

(b) The kidneys help maintain the balance of water and mineral salts (**1**) so sometimes more or less is left in the urine (**1**).

90. The role of ADH

1 (a) (i) Pituitary gland (**1**)

(ii) Collecting duct (**1**)

(iii) Concentration of urine increases (**1**) because ADH increases permeability of collecting duct to water (**1**).

2 (a) The volume of urine would be smaller (**1**) and the urine would be more concentrated (**1**).

(b) Water was lost from the body as sweat (due to running / hot day) (**1**), the pituitary (gland) responded by secreting more ADH into the blood stream (**1**) so that more water was reabsorbed from the kidney / nephron / collecting duct back into the blood (**1**) so the water content of the blood increased (**1**).

91. Kidney treatments

1 (a) Waste substances increase in concentration in the blood (**1**) and if not removed the person's life would be in danger (**1**).

(b) It separates blood from dialysis fluid (**1**) and allows small molecules to diffuse into the fluid (**1**).

(c) same (**1**), same (**1**), B higher (**1**), A higher (**1**)

(d) This maintains a concentration gradient across the partially permeable membrane (**1**) so urea continues to diffuse out of the blood (**1**).

2 The concentration in the blood would fall (**1**) because the concentration in the dialysis fluid would be lower so glucose would diffuse into fluid (**1**).

3 (a) A kidney is removed from a donor and put into the patient's body (**1**) and attached to their blood system (**1**).

(b) It may be harder to find a donor whose antigens match those on the patient's cells (**1**) to prevent rejection (**1**).

92. Extended response – Control and coordination

*Answer could include the following points:

- Cause of Type 1 diabetes: immune system has damaged insulin-secreting cells in pancreas, so no insulin produced.
- Cause of Type 2 diabetes: insulin-releasing cells may produce less insulin and target organs are resistant / less sensitive to insulin.
- Link risk of Type 2 diabetes with obesity / BMI / waist : hip ratio.
- Treat Type 1 diabetes by injecting insulin. Amount of insulin injected can be changed according to the blood glucose concentration.
- Treat Type 2 diabetes by diet (eating healthily and reduced sugar) and exercise.
- Treat more severe Type 2 diabetes with medicines to reduce the amount of glucose the liver releases or to make target organs more sensitive to insulin.

93. Exchanging materials

1 (a) Kidneys / nephrons (**1**) to maintain constant water level / osmoregulation (**1**)

(b) Kidneys / nephrons (**1**) urea is a toxic waste product (**1**)

2 In the lungs (**1**) oxygen is needed for respiration (**1**) carbon dioxide is a waste product (**1**)

3 (a) The surface of the small intestine is covered with villi (**1**). These help by increasing the surface area available for absorption (**1**).

(b) This makes the absorption of food molecules more efficient / effective (**1**) by reducing the distance that the molecules have to diffuse (**1**).

4 Four from: The flatworm is very flat and thin
(**1**) which means it has a large surface area :
volume ratio (**1**); the earthworm is cylindrical
so has smaller surface area : volume ratio
(**1**); every cell in the flatworm is close to the
surface (**1**); in the earthworm diffusion has to
happen over too great a distance / through too
many layers of cells (**1**).

94. Alveoli

1 (a) Oxygen diffuses from the air in alveoli
into the blood in capillaries (**1**). Carbon
dioxide diffuses from the blood into the
air (**1**).

(b) Millions of alveoli create a large surface
area for the diffusion of gases (**1**). Each
alveolus is closely associated with a
capillary (**1**). Their walls are one cell
thick (**1**). This minimises the diffusion
distance (**1**).

2 Maintains concentration gradient (**1**) which
maximises the rate of diffusion (**1**)

3 Three from: breathlessness / shortness of
breath / similar (**1**) less oxygen in blood
than normal (**1**) so less respiration / energy
(**1**); increased carbon dioxide concentration
reduces pH (**1**) which affects enzyme-
controlled reactions (**1**)

95. Rate of diffusion

1 (a) increase in surface area (**1**); shorter
diffusion distance (**1**); maintenance of a
high concentration gradient (**1**)

(b) surface area: alveoli in lungs (**1**);
diffusion distance: surfaces one cell thick
(**1**); concentration gradient: ventilation of
lungs / efficient blood supply (**1**)

2 (a) rate of diffusion would decrease (**1**) by
about 3 times (**1**)

(b) tiredness / fatigue (**1**); because less
oxygen for respiration (**1**)

(c) Reduced blood flow would reduce the
concentration gradient in the lungs (**1**)
and so less oxygen would be absorbed for
use in respiration (**1**).

96. Blood

1 (a) Red blood cells carry oxygen around
the body (**1**) and haemoglobin binds to
oxygen (**1**).

(b) Biconcave shape increases surface area
(**1**) to allow more diffusion of oxygen
(**1**). No nucleus (**1**) allows more space for
haemoglobin (**1**).

2 Dissolved substances such as glucose /
oxygen are transported to tissues (**1**). *Could
also say* Waste such as urea / carbon dioxide
(**1**) is transported away from tissues, to
kidneys / lungs (**1**).

3 Platelets respond to a wound by triggering
clotting process (**1**); clot blocks the wound (**1**)
and prevents pathogens from entering (**1**).

4 (a) Infection is caused by pathogen (**1**);
lymphocytes produce antibodies (**1**) that
stick to pathogens and destroy them (**1**).

(b) Phagocytes surround foreign cells (**1**) and
digest them (**1**).

97. Blood vessels

1 (a) An artery has thick walls (**1**). These walls
are composed of two types of fibres:
connective tissue (**1**) and elastic fibres (**1**).

(b) Wall stretches as blood pressure rises /
heart ventricles contract (**1**) and recoil
(*not contract!*) when blood pressure falls
/ heart ventricles relax (**1**).

2 (a) Thin walls / only one cell thick (**1**) run
close to almost every cell (**1**).

(b) faster diffusion of substances (**1**) because
short distance / large surface area (**1**)

3 (a) (i) Blood flows at low pressure (**1**) so no
need for elastic wall of arteries / need
wide tube in veins (**1**).

(ii) Muscles contract and press on veins
(**1**); blood pushed towards heart
because valves prevent flow the
wrong way (**1**).

(b) Veins have a thinner muscle wall than
arteries (**1**) so it is easier to get the needle
in (**1**). OR Veins contain blood under
lower pressure (**1**) so taking blood is more
controlled (**1**).

98. The heart

1 aorta – carries blood from heart to body (**1**);
pulmonary artery – carries blood from heart
to lungs (**1**); pulmonary vein – carries blood
from lungs to heart (**1**); vena cava – carries
blood from body to heart (**1**)

2 (a) because it acts as a pump (**1**) and muscles
contract to pump the blood (**1**)

(b) order of parts: (vena cava) right atrium,
right ventricle, pulmonary artery (lungs),
pulmonary vein, left atrium, left ventricle
(aorta) (names all correct for **2 marks**, 4
correct for **1 mark; additional mark** for
correct order)

3 (a) right ventricle (**1**); pumps blood to the
lungs / pulmonary artery (**1**)

(b) heart valve closes when ventricle
contracts (**1**); prevents backflow (**1**)

(c) has to pump harder (**1**) to get blood all
round body (**1**), not just to lungs (**1**)

99. Aerobic respiration

1 (a) oxygen and glucose (**2**)

(b) diffusion is the movement of substances
from high to low concentration (**1**)

2 (a) mitochondria (**1**)

(b) respiration is an exothermic process (**1**);
and transfers energy by heating (**1**)

(c) One from: to build larger molecules from
smaller ones (proteins from amino acids,
large carbohydrates from small sugars,
fats from fatty acids and glycerol) (**1**);
muscle contraction (**1**); active transport (**1**)

3 (a) glucose + oxygen → carbon dioxide +
water (**1**)

(b) capillaries (**1**)

4 (a) Respiration releases energy (**1**) so
that metabolic processes that keep the
organism alive can continue (**1**).

(b) Two from: Plants cannot use energy from
sunlight directly for metabolic processes
(**1**) so they need energy from respiration
for this purpose (**1**) during the day as well
as at night (**1**).

100. Anaerobic respiration

1 (a) Aerobic respiration releases more energy
(**1**) per molecule of glucose (**1**)

(b) The body needs energy more quickly
than aerobic respiration can supply (**1**);
it cannot get enough oxygen to respiring
cells (**1**).

2 (a) Heart rates will increase gradually /
remain low during early laps (**1**) and
increase rapidly during final sprint (**1**)
because energy demand is low at first
then increases significantly (**1**). You could
also say that adrenalin might increase
heart rate during early laps.

(b) Two from: to keep heart rate relatively
high (**1**) so that lactic acid is removed
from muscles (**1**); because oxygen is
needed to release energy needed to get rid
of lactic acid (**1**)

3 (a) Three from: oxygen consumption
increases during exercise (**1**) but reaches
a maximum value (**1**); no more oxygen
can be delivered for aerobic respiration
(**1**); increased energy needed comes from
anaerobic respiration (**1**)

(b) During exercise there is an increase in
the concentration of lactic acid (**1**); after
exercise, extra oxygen is needed to break
down lactic acid (**1**).

101. Rate of respiration

1 (a) maintains a constant temperature (**1**);
because temperature can affect enzymes /
change the rate of reaction (**1**)

(b) Absorbs carbon dioxide produced by the
seeds (**1**) so that this doesn't interfere
with the movement of the blob of water
(**1**).

(c) allows the pressure to be released
between experiments (**1**); so the blob of
water is pushed back to the start position
(**1**)

2 (a) Movement of the blob of water indicates
uptake of oxygen for use in respiration
(**1**), so measuring the movement of the
blob at intervals (**1**) allows the rate of
respiration to be calculated by dividing
distance moved by time taken (**1**).

(b) Use a water bath at a range of
temperatures (**1**); measure distance moved
by the blob over a particular time (**1**) and
repeat several times at each temperature
(**1**).

102. Changes in heart rate

1 (a) Stroke volume is the volume of blood
pumped from the heart in one beat (**1**).

(b) (i) cardiac output = stroke volume ×
heart rate = 60×75 (**1**) = 4500 (**1**)
cm^3 / minute (**1**)

(ii) Cardiac output increases (**1**); then
two from: cells need to respire faster
/ need more oxygen and glucose (**1**);
increased stroke volume / more blood
needed for respiring cells (**1**); so heart
rate must increase (**1**).

2 (a) $100 - 80 = 20$ (**1**); $20 / 80 \times 100 = 25\%$
(**1**)

(b) highest demand for oxygen / glucose /
respiration (**1**)

(c) Rearrange the equation to give stroke volume = cardiac output / heart rate (**1**) = 4000 / 50 = 80 (**1**) cm³.

103. Extended response – Exchange

*Answer could include the following points:

Outline of route:

- vena cava → right atrium → right ventricle → pulmonary artery → (capillaries in) lungs → pulmonary vein → left atrium → left ventricle → aorta → rest of the body / capillaries in the body → vena cava

Answer might also include:

- valves in heart / veins prevent backflow of blood
- deoxygenated blood enters / leaves right side
- oxygenated blood enters / leaves left side
- walls of left side of heart are thicker than right side

104. Ecosystems and abiotic factors

1 **1 mark** for each

Term — **Definition**

Community — A single living individual

Organism — All the living organisms and the non-living components in an area

Population — All the populations in an area

Ecosystem — All the organisms of the same species in an area

2 (a) south side = 323.5 (**1**); north side = 227.6 (**1**)

(b) She is correct (**1**) because the mean percentage cover is 39% on the south side compared with 4% on the north side / 10 times greater (**1**).

(c) Temperature (**1**) because it affects enzymes / rate of reactions (**1**) OR humidity (**1**) because water required for photosynthesis / other cell processes (**1**).

105. Biotic factors

1 (a) The living parts of an ecosystem (**1**).

(b) Two from: so that they can become the new alpha male (**1**); to gain fighting skills (**1**); to become stronger (**1**)

(c) Food can often be scarce in their habitat (**1**), so large groups need to split into different areas in order to find enough food (**1**).

2 The peacock has large, attractive tail feathers (**1**); it competes with other males for mates (**1**), so large showy tails are more attractive to female peahens (**1**). You could also suggest that the large tail feathers can be used to help scare away other male peacocks who may compete for females.

3 (a) The trees emerge through the canopy to get more light (**1**) for more photosynthesis (**1**).

(b) The trees have deep / extensive roots (**1**) to collect minerals (**1**).

106. Parasitism and mutualism

1 Both involve interdependence / the survival of one species is closely linked with another species (**1**); in mutualism both species benefit / win--win (**1**) but in parasitism one species is harmed / one species benefits at the expense of the other / win--lose (**1**)

2 (a) Fleas: obtain nutrients by sucking blood of the host (**1**); Animals: are harmed because they lose blood / nutrients / can catch disease from fleas (**1**)

(b) Parasitic (**1**) because fleas benefit and animals are harmed (**1**).

3 Cleaner fish get food by eating parasites from the skin of sharks (**1**). This helps the shark because it reduces the risk of the shark being harmed by the parasites (**1**).

4 scabies mite lives in the host and causes it harm (**1**); no benefit to the host (**1**)

107. Fieldwork techniques

1 (a) He could use a 1 m × 1 m quadrat (**1**), which he could throw at the flower bed to choose a random location (**1**).

(b) Using the same area means that his experiment is a fair test (**1**).

(c) He could look at more than one area each day (**1**) and take an average number of slugs (**1**).

2 (a) Find the total number of plants counted and the number of quadrats (**1**) and calculate a mean number of clover plants (**1**).

(b) total size of field = 100 × 65 = 6500 m² (**1**), so number of clover plants = 6500 × 7 (**1**) = 45 500 (**1**)

3 Place quadrats at regular intervals along the transect (**1**) and measure the percentage cover of broad-leaved plants in each quadrat (**1**). Record a named abiotic factor (light intensity / temperature) at each quadrat position (**1**).

108. Organisms and their environment

1 (a) Draw a line from the sea shore up the beach (at right angles to the sea line) (**1**) place quadrat at regular intervals along the line (**1**) count the limpets in each quadrat area (**1**).

(b) (10 + 8 + 9) / 3 (**1**) = 9 (**1**)

(c) The number of limpets goes down as you travel further from the sea (**1**); this decrease is linear with the distance / it drops by 4 limpets for every 0.5 m distance travelled (**1**); limpets are more likely to survive if they live nearer the seashore (**1**).

2 (a) Instead of placing a quadrat every 2 m, the scientist could place a quadrat every 0.5 m / 1 m / smaller distance (**1**) and use a smaller (**1**) quadrat than before.

(b) Two from: the number of bluebells increases to a maximum around 8 m into the wood (**1**); less light is available for photosynthesis (**1**) and fewer nutrients / water available deeper in the wood where there are more trees (**1**).

109. Energy transfer between trophic levels

1 (a) grass (**1**)

(b) four (**1**)

(c) Because there is not enough biomass in the top level (**1**) to provide the energy needed by another level (**1**).

2 (a)

Organism	Energy at each trophic level (J)	Number of organisms	Mass of each organism (kg)	Biomass at each trophic level (kg)
Producers	7550	10 000	0.25	2500
Herbivores	640	200	2.5	500
Carnivores	53	10	20	200

(**1 mark** for all correct answers)

(b) correctly shaped pyramid with producers at bottom, then herbivores, then carnivores (**1**) horizontal width is drawn to scale (**1**)

(c) (53 ÷ 640) × 100 (**1**) = 8.3% (**1**)

(d) Some of the energy from respiration (**1**) is transferred as heat to the environment (**1**).

110. Human effects on ecosystems

1 Advantage – reduces fishing of wild fish. (**1**)

Disadvantage – one of: the waste can pollute the local area changing conditions so that some local species die out (**1**); diseases from the farmed fish (e.g. lice) can spread to wild fish and kill them (**1**)

2 Advantage – one of: may provide food for native species (**1**) may increase biodiversity (**1**)

Disadvantage – one of: may reproduce rapidly as they have no natural predators in the new area (**1**); may out-compete native species for food or other resources (**1**)

3 (a) 145 – 15 = 130; (130 / 15) × 100 (**1**) = 867% (**1**)

(b) increasing population (**1**); more food / crops needed (**1**)

(c) Excess fertiliser can be leached / washed into rivers / lakes (**1**), causing eutrophication (**1**).

111. Biodiversity

1 (a) replanting forests where they have been destroyed (**1**)

(b) Two from: restores habitat for endangered species (**1**); reduces carbon dioxide concentration in the air (as trees photosynthesise) (**1**); reduces the effects of soil erosion (**1**); reduces range of temperature variation (**1**)

2 Some species are valuable to humans (**1**) because they are a source of new drugs / are wild varieties of crop plants / source of genes (**1**).

3 The numbers of trees will increase because there are fewer deer to eat them. **(1)** This means that there will be more food for birds / bears / rabbits / insects. **(1)** There will be more rabbits because there are fewer coyotes to kill / eat them. **(1)** If there are more rabbits, there will be more food for coyotes / hawks / predators. **(1)** More trees also mean that there will be more habitats for birds and less soil erosion. **(1)**

112. Food security

1 (a) access to a reliable **and** adequate food supply **(1)**

(b) Greater numbers mean more food is needed **(1)**; as people become better off there is greater demand for meat and fish **(1)**.

(c) being able to take what you need to live now **(1)** without damaging the supply of resources in the future **(1)**

2 (a) It would reduce the amount of land for growing food crops **(1)** and so food would become more expensive / poor people could not afford to buy food **(1)**.

(b) Two from: growing crops for biodiesel would need use of more fertilisers / pesticides **(1)** or might increase pollution **(1)**; could cause deforestation to create more land for growing crops for biodiesel **(1)**; could introduce new pests / diseases **(1)**

3 Three from: 7 kg of grain will feed more people than 1 kg of beef **(1)**; more land will be required to grow grain for animal feed **(1)** and this could result in deforestation **(1)**; intensive farming creates more pollution **(1)** and could increase spread of pests / diseases **(1)**

113. The carbon cycle

1 (a) photosynthesis **(1)**

(b) respiration **(1)**

(c) combustion **(1)**

(d) decomposition **(1)**

2 Microorganisms are decomposers **(1)**; they convert complex carbon-containing molecules into carbon dioxide (by respiration), which is released into the atmosphere **(1)**.

3 (a) Fish carry out respiration **(1)**; respiration releases carbon dioxide into the water **(1)**; plants absorb the carbon dioxide **(1)**, which is used in photosynthesis **(1)**.

(b) Any three from: if there are not enough fish / snails / aquatic animals in the tank there will not be enough carbon dioxide **(1)**, so there is less photosynthesis by plants **(1)**, so less oxygen is released for fish / snails / aquatic animals **(1)**; less food for animals as fewer plants **(1)**; plants and fish / snails / aquatic animals die **(1)**

114. The water cycle

1 (a) Three from:
Evaporation from land **(1)** sea **(1)** and transpiration from plants **(1)** animal sweat **(1)** animal breath **(1)**

(b) Water vapour condenses to form clouds **(1)**; water cools to form precipitation / rain / snow **(1)** that returns the water to Earth **(1)**

2 A lot of water evaporates from a golf course so this will lead to more water in the atmosphere **(1)**; water levels fall in the river as water is removed for watering the golf course **(1)** so animals or plants living in the river might die **(1)**.

3 Advantage: sea water is made potable / safe to drink **(1)**

Disadvantage: needs a lot of energy / fuel / it is expensive **(1)**

115. The nitrogen cycle

1 C **(1)**

2 (a) Nitrogen fixation by soil bacteria **(1)**.

(b) Three from: nitrates are absorbed by roots **(1)** by active transport **(1)** because plants need nitrogen for making amino acids / proteins **(1)** but can only take in nitrogen in the form of nitrate / ammonium (ions / salts) **(1)**

(c) Amount of nitrate in soil is reduced **(1)** because bacteria convert nitrates in the soil into nitrogen gas in the air **(1)**

3 (a) Three from: Plants such as clover have nitrogen-fixing bacteria in their roots **(1)** so they can be grown and ploughed back into the soil **(1)**; they are decomposed **(1)** to add nitrates **(1)**.

116. Pollution indicators

1 (a) When sewage enters the river, the amount of oxygen drops **(1)** and then rises further down the river **(1)**.

(b) two from: stonefly larvae **(1)**; mayfly **(1)**; caddisfly **(1)**

(c) Four from: nitrates in sewage will cause plants / algae to grow **(1)**; algae block out sunlight to plants below **(1)**; no photosynthesis so plants die **(1)**; bacteria decompose dead organisms, using up oxygen and leaving less oxygen for fish / aquatic animals **(1)**; fish / animals die **(1)**

2 Four from: A high number of certain lichen species in a certain area **(1)** indicate high levels of sulfur dioxide **(1)**; presence of blackspot fungus **(1)** indicates low / no sulfur-dioxide pollution **(1)**; but other biotic / abiotic factors can affect indicator species **(1)**

117. Decay

1 (a) The conditions needed are oxygen, water / moisture **(1)** and warm temperatures **(1)**.

(b) The sun will increase the temperature in the heap **(1)** and the enzymes will work faster **(1)**.

(c) To increase the rate of decay **(1)** because microorganisms need moisture **(1)**.

2 (a) Drying: decomposer microorganisms need water for cellular processes **(1)**. Salting: causes water to move out of bacterial cells by osmosis **(1)** so there is not enough water in the cells for them to grow **(1)**.

(b) Refrigeration: cold temperatures mean microorganisms grow more slowly **(1)** because enzyme activity is reduced at low temperatures **(1)**. Packing in nitrogen: replaces oxygen **(1)** and microorganisms need oxygen to respire **(1)**.

118. Extended response – Ecosystems and material cycles

* Answer could include the following points:

- Fish farming can reduce biodiversity by introducing just one species.
- Waste and diseases can affect wild populations.
- Introduction of non-native species might lead to competition with native species.
- Reduces fishing of wild fish.
- Fertilisers can cause eutrophication leading to loss of biodiversity in nearby water.
- You could also talk about conservation, reforestation, captive breeding.

119. Timed Test 1

1 (a) Add hydrogen peroxide to each test tube **(1)**; add tissue to each tube **(1)**; measure height of foam after 30 seconds / same length of time **(1)**.

(b) Two from: keep temperature constant / use a water bath **(1)**; use same volume / concentration of hydrogen peroxide **(1)**; use same size / weight / amount of tissue each time **(1)**; grind / liquidise tissues so they have same surface area **(1)**; use a buffer to maintain pH **(1)**

(c) Collect the gas in a gas syringe to measure its volume **(1)**.

2 (a) communicable **(1)** because it is caused by a bacterium / pathogen **(1)**

(b) $((3.20 - 1.45) / 3.20) \times 100$ **(1)** $= 54.7\%$ **(1)**

(c) The number of bacteria was reduced **(1)** and the man felt better **(1)** and antibiotics do not kill viruses **(1)**.

(d) Three from: The number of bacteria has only been reduced by about half **(1)** so they could grow again / the disease could come back if he stopped early **(1)**; the remaining bacteria are more likely to be resistant to the antibiotic **(1)** so there is a risk of increasing antibiotic-resistant bacteria **(1)**.

3 (a) restriction endonuclease **(1)**

(b) 2, 1, 3, 5, 4: all correct **(2)**, 1 mistake **(1)**

(c) If the same restriction enzyme is used to make the fragment and cut the plasmid **(1)** the sticky ends will match **(1)** and DNA ligase can be used to make a complete recombinant plasmid **(1)**.

(d) advantage: Crop damage is reduced so yield should increase / less chemical insecticide is needed **(1)**. Disadvantage: Bt broad beans will be more expensive / insect pests may become resistant / Bt gene may transfer to wild plants **(1)**.

4 (a) nucleus **(1)**

(b) (i) makes / produces / synthesises proteins / enzymes **(1)**

(ii) Two from: plant cell has vacuole **(1)** / chloroplasts **(1)** / nucleus **(1)** / is much larger **(1)**

(c) $59\,000 \div 500$ **(1)** $=118$ **(1)**; third mark for 1.2×10^2

(d) bacterial cell = electron microscope, plant cell = light microscope **(1)** because light microscope has maximum magnification of approx. $\times 1500$ **(1)**

143

5 (a) Measure weight and height (**1**) every year / at regular intervals (**1**) and plot the results on the chart (**1**).

(b) (i) all points correctly plotted to ±0.5 division, (**1**) per set of points (height and weight for both boys)

(ii) Boy A is average / below average height (**1**) but in the upper range for weight / overweight (**1**); boy B is above average height (**1**) and average weight / possibly underweight for his height (**1**).

6 (a) B (**1**)

(b) (i) UAGCCAGAUGGC (**1**)

(ii) 990 ÷ 3 = 330 (**1**) (each amino acid is coded for by 3 bases)

(c) Being able to determine the base sequence of an individual (**1**) means that doctors know if they have an increased risk of testicular cancer (**1**) so they can be monitored more closely (**1**).

7 (a) (i) incubate the plates at 25 °C / don't completely seal the Petri dishes (**1**)

(ii) Two from: sterilise agar before use (**1**); sterilise Petri dishes before use (**1**); sterilise / flame inoculating loops (**1**); seal the lid on the dish with tape (**1**); make sure all surfaces are clean / disinfected before use (**1**)

(b) (i) A = 2.2 (**1**); B = 53.6 (**1**)

(ii) Antibiotic A is more effective than antibiotic B (**1**) but neither antibiotic completely kills / prevents growth of bacteria (**1**).

8 (a) There is a positive correlation between alcohol consumption and risk of liver disease / as alcohol consumption increases risk of liver disease also increases (**1**); the risk increases much more with alcohol consumption greater than 50 g per day (**1**) because ethanol is poisonous, particularly to liver cells (**1**).

(b) The risk of liver disease is higher for women than for men at all levels of alcohol consumption (**1**) and the risk increases even more steeply above 40 g per day (**1**).

(c) A (**1**)

9 (a)

Statement	Mitosis only	Meiosis only	Both mitosis and meiosis
used for growth and replacement of cells	✓		
used for production of gametes		✓	
before the parent cell divides each chromosome is copied			✓
produces genetically identical cells	✓		
halves the chromosome number		✓	

1 mark for each correct tick

(b) (i) advantage: offspring are genetically different / source of variation that is the basis of natural selection / if the environment changes some individuals may survive (**1**) disadvantage: need to find a mate / requires more time and energy (**1**)

(ii) advantage: offspring are genetically identical to parent so if parent is well adapted to environment offspring will be too / only one parent so no need to find a mate / reproductive cycle is faster (**1**) disadvantage: no variation in the population / if environment changes all may die (**1**)

(c) (i) ring round 'X' and 'Y' of parent 1 (**1**)

(ii) completion of Punnett square (**1**); half the offspring are female (XX) (**1**) so chance is 50% / 1 in 2 (**1**)

parent 1

		X	Y
parent 2	**X**	X X	X Y
	X	X X	X Y

(d) *Answer could include the following points:

- Sexual reproduction is a source of variation.
- Some individuals will be better adapted to their environment, particularly if conditions change;
- individuals with advantageous variations in characteristics will pass their genes on;
- those with not such well adapted variations less likely to survive;
- more individuals will have the advantageous variations in subsequent generations.

10 (a) D (**1**)

(b) (i) synapse (**1**)

(ii) Three from: nerve impulse reaches the axon terminal (**1**); neurotransmitter substance released into the gap (**1**); this is detected by the next neurone (**1**) which generates a new impulse (**1**)

(c) (i) It would cause loss of feeling in the hand (**1**) because impulses from sensory receptors would not be passed on (**1**).

(ii) Damage to the spine would cause paralysis (**1**) as well as loss of feeling (**1**) because motor neurones would also be damaged (**1**).

(d) (i) The extent of paralysis will indicate the position of the injury (**1**); more detailed information could come from CT scan (**1**) or PET scan (**1**).

(ii) *Answer could include the following points:

Stem cells are unspecialised cells; they can divide and produce differentiated cells; embryonic stem cells are taken from embryos at a very early stage of division; they can be used to repair damaged nervous tissue; they are easy to extract from embryos and can produce any type of cell; however embryos are destroyed and some people think embryos have a right to life; there is a danger of rejection because the transplanted tissue is seen as foreign; adult stem cells do not require destruction of embryos and will not be rejected (if taken from the person to be treated); but they may not be suitable for treating damaged nerve tissue; stem cell therapy may increase risk of cancer.

125. Timed Test 2

1 (a) (i) thermometer / temperature probe (**1**)

(ii) Two from: heat might be lost before the burning food is placed under the tube (**1**); the food might be held too far away from the tube / at different distances (**1**); draughts might mean heat was lost / did not heat the water (**1**); heat is absorbed by the test tube as well as the water (**1**)

(iii) energy transferred = 500 × 4.2 × 60.8 (**1**) = 127 680 J = 127.7 kJ (**1**); energy content = 127.7 / 8.3 = 15.4 kJ / g (**1**)

(b) (i) biuret test (**1**)

(ii) pale purple / lilac (**1**)

(c) Two from: because more protein will increase the nutritional value of the rice (**1**); as populations grow the demand for protein increases (**1**); increased protein content will mean more food can be grown in the same amount of land (**1**) or without using more fertiliser / pesticide (**1**)

2 (a) (i) pituitary gland (**1**)

(ii) It stimulates growth and maturation of follicles (**1**).

(b) High levels of oestrogen stimulate release of more luteinising hormone (LH) (**1**) and the LH surge triggers ovulation (**1**).

(c) (i) Clomifene stimulates production of FSH and LH (**1**) so women who don't normally produce enough of these hormones (**1**) will ovulate normally with the help of clomifene (**1**).

(ii) Clomifene stimulates the maturation of many follicles / eggs (**1**) so that there are more eggs available for fertilisation in vitro / IVF (**1**).

3 (a) (i) 60–65 beats per minute (**1**)

(ii) 7.30 – 8.00 am (**1**)

(iii) Heart rate increases to a peak of about 80 as he walked uphill (**1**) and then falls again shortly before the main peak as he rested (**1**).

(iv) because his muscles were starting to respire anaerobically (**1**) and so they tired more quickly (**1**)

(v) because extra oxygen is needed to replace oxygen used in the exercise (**1**) and to oxidise lactic acid produced (**1**)

(b) (i) person A = 95 × 52 = 4940 **(1)**;
person B = 58 × 72 = 4176 **(1)**; units
= cm³ per min **(1)**

(ii) person A **(1)** because they had a lower
resting heart rate / higher cardiac
output / higher stroke volume **(1)**

4 (a) (i) Y = capillary wall **(1)**; Z = wall of
alveolus **(1)**

(ii) diffusion **(1)**

(iii) carbon dioxide **(1)**

(iv) in red blood cells / bound to
haemoglobin **(1)**

(b) (i) 480 × 0.15 = 72 m² **(1)**

(ii) Rate of diffusion is proportional to
surface area **(1)**, so the rate of gas
exchange will have increased **(1)**,
which will allow the athlete to absorb
more oxygen for use in aerobic
respiration **(1)**.

5 (a) (i) A parasite obtains food / nutrients
/ water / shelter / benefit from the
host **(1)** and harms the host **(1)** but
an epiphyte does not harm the host /
obtains its nutrients / water from the
surroundings **(1)**.

(ii) Nitrogen-fixing bacteria are protected
by and get food from the plant **(1)** and
the legume gets nitrogen compounds
it needs from the bacteria **(1)** so each
organism benefits **(1)**.

(b) (i) One from: large leaves to absorb as
much light as possible **(1)**; drip tips
on the leaves so water runs off **(1)**

(ii) Two from: waxy cuticle **(1)**; stomata
sunk in pits to reduce water loss **(1)**;
leaf hairs to trap moist air round
stomata **(1)**; rolled leaves to reduce
air movement around stomata **(1)**

6 (a) (i) adrenal gland **(1)**

(ii) Adrenalin is secreted into the blood
(1) and transported around the body
in the blood **(1)**.

(iii) Two from: hormones are secreted
into the blood, nerve impulses travel
along nerves **(1)**; hormones have a
long-lived effect, nerves have a short-
term effect **(1)**; nerve impulses work
quickly, hormones take longer to
work **(1)**; hormones can have effects
on many different parts of the body,
nerves act on specific organs **(1)**

(b) (i) glucagon **(1)**

(ii) diabetes **(1)**

(c) *Answer could include the following
points:

Negative feedback operates when a
factor rises above normal. A corrective
mechanism brings the factor back to the
normal level. If the factor falls below
normal, a different corrective mechanism
brings the factor back to normal.
Thermoregulation is controlled by the
hypothalamus. If body temperature rises
this is detected by the hypothalamus and
leads to:

vasodilation so more blood flows near the
surface of the skin; sweat glands release
more sweat onto the skin surface to
evaporate; sebaceous glands produce oil
to help sweat spread out and so evaporate
more easily; more heat transferred to
environment so body temperature falls.

If the body temperature falls this is
detected by the hypothalamus and leads
to:

vasoconstriction so less blood flows near
the surface of the skin; sweat glands stop
production of sweat; body hairs raised by
erector muscles; shivering, to generate
heat; less heat transferred to environment
so body temperature rises.

7 (a) (i) Two from: light intensity **(1)**; water
availability **(1)**; mineral ions in the
soil **(1)**; temperature **(1)**

(ii) two from: predation / grazing by
animals **(1)**; competition for light and
space **(1)**; competition for water and
nutrients **(1)**

(b) Use a belt transect **(1)**; place quadrats at
regular intervals alongside the path **(1)**;
count the number of each different plant
species in each quadrat / calculate the
percentage cover of each different plant
species **(1)**; present the results as a table /
graph **(1)**.

8 (a) (i) ((5.4 − 5.2) ÷ 5.2) × 100 = +3.8%;
((5.6 − 5.6) ÷ 5.6) × 100 = 0.0%;
((5.4 − 5.6) ÷ 5.6) × 100 = −3.6%;
((4.6 − 5.0) ÷ 5.0) × 100 = −8.0%;
((4.2 − 5.2) ÷ 5.2) × 100 = −19.2%;
all correct = **3 marks**, 4 correct =
2 marks, 3 correct = **1 mark**.

(ii) One from: same shape / same size /
same temperature / same time left in
the solution **(1)**

(iii) One from: repeat and calculate
mean values / use a smaller
range of concentrations (near the
0% concentration) / use more
concentrations in the same range / use
a ±0.01 g balance **(1)**

(iv) The solute concentration of the potato
cells must have been 0.5 mol per dm³
(1) because there was no change in
mass / movement of water in or out
(1).

(b) Dialysis tubing is partially permeable **(1)**
so that urea diffuses out of the blood into
the fluid **(1)** but dialysis fluid contains the
same concentration of useful substances
/ glucose / mineral ions **(1)** so diffusion
restores the normal concentration of
dissolved substances in the blood **(1)**.

9 (a) (i) B **(1)**

(ii) E sieve tube **(1)**; F sieve plate **(1)**

(iii) Three from: mitochondria supply
energy **(1)** from respiration **(1)** which
is used to move sucrose in and out **(1)**
by active transport **(1)**.

(b) (i) light intensity / carbon dioxide
concentration / amount or mass of
chlorophyll **(1)**

(ii) Rate increases because reactions
happen faster at high temperatures
(1) and enzymes work faster **(1)**
but it decreases again because high
temperatures denature enzymes **(1)**.

10 (a) B **(1)**

(b) ((12 − 2) ÷ 2) × 100 = 500% **(1)**

(c) *Answer could include the following
points:

Disadvantages of salmon farming:
uneaten food can cause pollution;
salmon produce waste and this can cause
pollution; this can lead to local species
dying out; diseases (e.g. sea lice) from the
farmed salmon can transfer to wild fish
and harm or even kill them.

Advantages of salmon farming: provides
employment; reduces fishing of wild fish.

Advantages of mussel farming: no food
is added to the water, so no uneaten food
to cause pollution; mussels remove waste
from the water.

Disadvantages of both: reduces
biodiversity; can spoil the landscape.

Your own notes

Your own notes

Your own notes

Your own notes

Published by Pearson Education Limited, 80 Strand, London, WC2R 0RL.

www.pearsonschoolsandfecolleges.co.uk

Copies of official specifications for all Pearson qualifications may be found on the website:
qualifications.pearson.com

Text © Pearson Education Limited 2017
Typeset and produced by Phoenix Photosetting
Illustrated by Phoenix Photosetting
Cover illustration by Miriam Sturdee

The rights of Stephen Hoare to be identified as author of this work has been asserted by him in accordance with the Copyright, Designs and Patents Act 1988.

First published 2017

20 19 18

10 9 8 7 6 5 4 3 2

British Library Cataloguing in Publication Data
A catalogue record for this book is available from the British Library
ISBN 978 1 292 13176 4

Acknowledgements
The publishers are grateful to Sue Kearsey, Allison Court and Nigel Saunders for their help and advice with this book.

The author and publisher would like to thank the following individuals and organisations for permission to reproduce copyright material:

Photographs
(Key: b-bottom; c-centre; l-left; r-right; t-top)

Science Photo Library Ltd: BioPhoto Associates 3, Steve Gschmeissner 6, 15

All other images © Pearson Education

Figures
Figure on page 55 adapted from data from the National Health Service of the United Kingdom, http://www.pbs.org/wgbh/nova/body/autism-vaccine-myth.html; figure on page 61 adapted from 'Coronary Heart Disease prevalence for males and females by age and ethnicity (self-assigned and Origins-assigned) in West Midlands' from *Key Health Data 2009/10*, figure 11.1b, copyright © Public Health, Epidemiology and Biostatistics Unit, School of Health and Population Sciences, University of Birmingham http://medweb4.bham.ac.uk/websites/key_health_data/2009/figures/ch_11/fig_11.01b.htm; figure on page 110 from Revise AQA GCSE Additional Science Revision Workbook Higher by Iain Brand and Mike O'Neill, Pearson Education Ltd, 2013, reproduced with permission; figure on page 121 adapted from '2 to 20 years: Boys Stature-for-age and Weight-for-age percentiles', published 30/05/2000, modified 21/11/2000, developed by the National Center for Health Statistics in collaboration with the National Center for Chronic Disease Prevention and Health Promotion, https://www.healthychildren.org/Documents/tips-tools/Growth%20Charts/2_20_years_boys_stature_weight.pdf; figure on page 123 adapted from http://pubs.niaaa.nih.gov/publications/arh27-3/images/mann1.gif, copyright © NIAAA, National Institute on Alcohol Abuse and Alcoholism; figure on page 125 adapted from http://www.rsc.org/learn-chemistry/resource/res00000397/energy-values-of-food?cmpid=CMP00005022, copyright © Nuffield Foundation and the Royal Society of Chemistry, updated October 2015, reproduced with permission; figure on page 129 adapted from 'Salmon lice – impact on wild salmonids and salmon aquaculture' from *Journal of Fish Diseases* by O. Torrissen, S. Jones, F. Asche, A. Guttormsen, O. T. Skilbrei, F. Nilsen, T. E. Horsberg, and D. Jackson, Vol 36 (3), pp.171–194, March 2013, copyright © 2013 Blackwell Publishing Ltd.

Notes from the publisher
1. In order to ensure that this resource offers high-quality support for the associated Pearson qualification, it has been through a review process by the awarding body. This process confirms that this resource fully covers the teaching and learning content of the specification or part of a specification at which it is aimed. It also confirms that it demonstrates an appropriate balance between the development of subject skills, knowledge and understanding, in addition to preparation for assessment.

Endorsement does not cover any guidance on assessment activities or processes (e.g. practice questions or advice on how to answer assessment questions), included in the resource nor does it prescribe any particular approach to the teaching or delivery of a related course.

While the publishers have made every attempt to ensure that advice on the qualification and its assessment is accurate, the official specification and associated assessment guidance materials are the only authoritative source of information and should always be referred to for definitive guidance.

Pearson examiners have not contributed to any sections in this resource relevant to examination papers for which they have responsibility.

Examiners will not use endorsed resources as a source of material for any assessment set by Pearson.

Endorsement of a resource does not mean that the resource is required to achieve this Pearson qualification, nor does it mean that it is the only suitable material available to support the qualification, and any resource lists produced by the awarding body shall include this and other appropriate resources.

2. Pearson has robust editorial processes, including answer and fact checks, to ensure the accuracy of the content in this publication, and every effort is made to ensure this publication is free of errors. We are, however, only human, and occasionally errors do occur. Pearson is not liable for any misunderstandings that arise as a result of errors in this publication, but it is our priority to ensure that the content is accurate. If you spot an error, please do contact us at resourcescorrections@pearson.com so we can make sure it is corrected.